Simple Tomato Growing

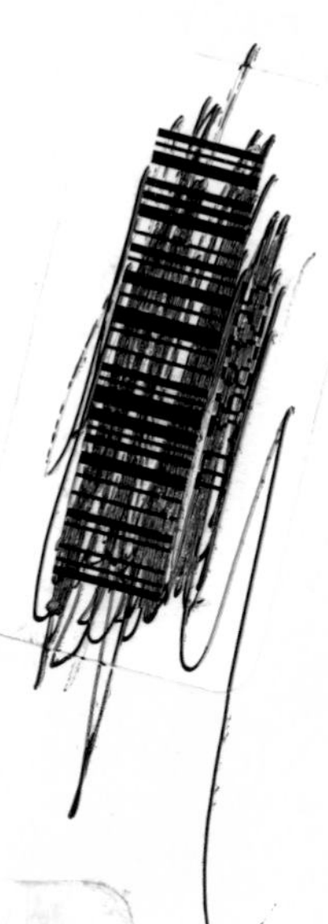

Other Concorde Books

Gardening for Beginners
Simple Greenhouse Gardening
Making and Planning a Small Garden
Simple Vegetable Growing
Practical Pruning
Shrubs and Decorative Evergreens
House Plants
Hardy Plants for Small Gardens
The Perfect Lawn
Simple Plant Propagation
How to Grow and Use Herbs

Simple Tomato Growing

Ian Walls

Ward Lock Limited · London

Cased ISBN 0 7063 1907 9

Paperback ISBN 0 7063 1905 2

First published in Great Britain 1975
by Ward Lock Limited, 116 Baker
Street, London, W1M 2BB

Reprinted 1976

Text filmset in 11 pt Times

Printed and bound in Great Britain
by Cox & Wyman Limited,
London, Fakenham and Reading

Contents

Preface

Tomatoes are not only the most colourful of all vegetables but also the most rewarding to grow. While you can go out and buy tomatoes from your local shop or supermarket at any time of year, those that you have grown yourself have far more flavour (at least most people who grow their own claim that this is so), and you can grow many varieties in your garden that you cannot buy in the shops. The reason is simple: commercial growers have to concentrate on varieties which are heavy croppers giving maximum profit, varieties which pack well, travel without bruising and last well in the shops. Many of the most tasty varieties do not meet these criteria. These are the ones to concentrate on in your own garden.

This brief book is written for every gardener who wants to grow tomatoes and who wants to grow them well but does not know how best to go about it.

It tells you how to grow tomatoes in a greenhouse, a polythene shelter, a porch, on a window sill, or out of doors in a sunny corner.

Basically all you need to know in order to succeed with tomatoes is what to do, why you do it that way, and when to do it. Armed with that information, success is yours. The purpose of this book is to give you just that information.

1

The Tomato and its Background

All cultivated plants are forms of wild plants that have been improved to a greater or lesser degree by man, by careful selecting and breeding. The large, luscious fruits of the strawberry, for example, have been developed from the small, hard and often bitter fruits of the common wild strawberry.

The tomato has been improved rather more than most plants—literally almost out of recognition. The wild plant is a native of Peru, where it spreads and sprawls about on the ground like many of the weeds which are a nuisance to European gardeners. Several different forms are known in the wild, some of which have larger fruits than others, and some of which have a more robust habit, though most have rather small, sharp fruits.

The tomato was introduced to Europe in the mid-sixteenth century and to Britain in 1595. For a long time they were grown only for their decorative fruits which were thought to be poisonous. The first reference to tomatoes being grown for culinary use was made in a book published in 1781. Tomatoes were originally known in Europe as 'love apples' and were claimed to have aphrodisiac qualities! It was not until the latter part of the nineteenth century that the tomato was grown widely in Europe, mainly in greenhouses, and out of doors in America.

Since the discovery was made that tomato fruits are edible, plant breeders have, by a process of selection and interbreeding, developed the tomato plants we know today.

In America and the Mediterranean countries the large 'meaty' tomatoes are the most popular, but people throughout most of northern Europe tend to prefer the small round or oval tomato, thought by many to be of higher quality and more pleasant to eat than the large tomato.

Yellow, red, orange or striped tomatoes, tomatoes with large, small, round or long, smooth or corrugated fruits are all available. One enterprising firm in England will even provide you with a do-it-yourself tomato breeding kit with which you can have endless fun breeding your own varieties.

The climate of the massive coastal plain in Peru where the tomato comes from is much warmer on average than that of Europe. Consequently, tomatoes are seldom really successful if grown entirely out of doors. Some extra heat is needed to germinate the seed, and usually to see the young tomato plants through their first few weeks until they are robust enough to plant in the open. Alternatively, the artificial climate which can be provided in a greenhouse or conservatory in temperate countries such as Britain is ideal for tomato growing. If the greenhouse can be heated during the spring and autumn months the plants will yield better crops than they would without this extra heat or out of doors.

Natural light is important for the growing of any plant, but particularly important for growing tomatoes. The part of South America where the tomato comes from has a constantly high light level. Despite all the changes which the plant breeder has brought about in the tomato, it still likes plenty of light and is not too happy during the winter and early spring in Britain. However, artificial light can be used to make up for the poor natural light during the dull days early in the year.

Although a greenhouse in which some heat is available is the best place in which to grow tomatoes, cold greenhouses, cloches, porches or window sills can be used to produce later, but still rewarding, crops. Out of doors results can be good, especially in a warm summer, and particularly if robust, well-started plants are used.

2

Where to Grow Tomatoes

Whatever you do in life, the greater the effort you put into it, the greater your reward will be, and this is just as true of gardening as it is of any other activity. When it comes to growing tomatoes the first thing to appreciate is that they are demanding plants. Unless you are prepared to spend quite a lot of time and effort on them the results will be so poor they will hardly be worth growing at all.

Tomatoes, coming from a warmer climate than ours, are not well adapted to growing out of doors in Britain and in countries with similar climates. To start with they are not hardy: the slightest touch of frost will kill their foliage, while a more severe frost will kill the plant completely. This is a very real problem when you realize that throughout most of the country the last of the spring frosts occur as late as late May, and the first of the winter frosts as early as September. This leaves only three frost-free months, and that is not enough for a tomato plant to reach a sufficiently advanced stage of growth to set mature, edible fruits. At the Royal Horticultural Society's garden at Wisley near Woking some frost is recorded during every month of the year, and this situation is not uncommon.

There are other factors besides frost which make it difficult to grow tomatoes really successfully out of doors in climates like that of north-western Europe, including Britain. Tomatoes like sun and warmth, but all too often the summers are dull,

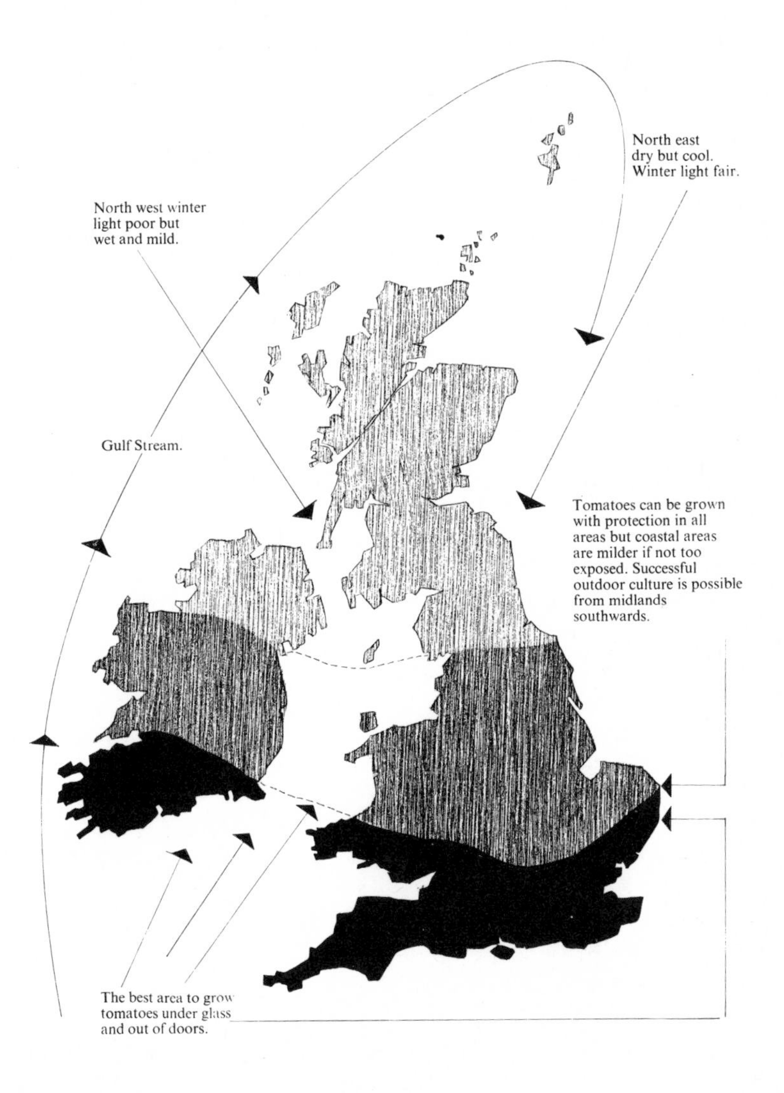

A map showing the climatic zones within the UK

grey and rather chilly. Tomatoes also like a constant but moderate supply of moisture at the roots, and dislike extremes of drought and waterlogging, especially when these extremes alternate, as they often do out of doors.

For these reasons, and for several others which will be dealt with later, by far the most satisfactory way of growing tomatoes is in a greenhouse or some similar structure which will protect the plants from the vagaries and extremes of the weather. The great advantage of this is that it not only enables you to provide the plants with ideal growing conditions, but also greatly extends the growing season, thereby ensuring that the plants fruit successfully.

Although in theory it would be great fun to grow tomatoes out of season, this is very seldom done, even by commercial growers. There are sound practical reasons for this. Apart from the very considerable expense that would be incurred in heating the greenhouse to the required temperature in winter, there is the far greater problem of providing the plants with sufficient light. Tomatoes need all the light they can get, not only to ripen their fruit, but simply to grow. Few of us realize just how much weaker the sun is in winter than in summer. To grow tomatoes out of season one would have to provide artificial light all winter, often well into the night. Apart from the expense of this, your neighbours might well complain that your gaily lit greenhouse kept them awake at night, and there could be by-laws which prevent you from lighting the greenhouse at night anyway. On a small scale, however, a terraria complete with artificial lighting can be used with startling results. These are items of equipment limited in size and more useful during the early stages of raising the plants, though there are some dwarf varieties which can be grown to maturity in large terraria.

Obviously the most sensible way of growing tomatoes is to wait until natural light conditions improve and the weather is warmer, especially with the high cost of fuel today. A practical growing timetable will be given later.

Bright sunlight comes mainly from the south in the northern hemisphere. It is therefore important to select a site for the greenhouse or other structure which is unobstructed to the south. Trees, buildings or even tall fences can greatly restrict sunlight and these should never be allowed to come between the sun and the greenhouse. This is especially so during the vital early spring period when the sun is very low in the sky.

Another important point, and one which is often overlooked, is the reflective property of the glass on a greenhouse. Although it usually seems that glass allows all the light reaching it to pass through, in fact a proportion of the light is always reflected away. The amount of light deflected will depend upon the angle at which the sun's rays hit the glass. If they hit the glass at right angles, the highest possible proportion of light will pass through. The further they deviate from a right angle the higher the proportion that will be deflected away. The diagram illustrates this point very clearly.

Because of this it is important to run greenhouses of the normal rectangular shape with their long axis lined up east/west, since this allows the whole vertical or slightly sloping side to face advantageously to the south. Greenhouses sited with their long axis running north/south tend to be darker in the winter and early spring. This matters less once the sun starts to get higher in the sky. After about April or May it really does not matter too much which way the greenhouse runs and almost any light-admitting structure or protecting device will grow tomatoes. It is in the early stages of the tomatoes' growth that this could be crucial.

If you are using a porch, conservatory, light window or frame the same applies, but you will get still better results if they face due south or south-west.

Because of these factors it is worth surveying briefly the places and types of structures in which tomatoes can be grown, and how to get the best results from them.

GREENHOUSES

Any greenhouse can be used to grow tomatoes, but the more sunlight it lets in, the better the results you will achieve.

Greenhouses can be made of alloy or wooden superstructures with different sizes of glass panes. Glass can either reach to the ground or there can be a base 'wall' of brick, wood, asbestos or metal.

The purpose of any greenhouse or similar structure is to admit as much light and heat as possible available from the

The use of trees to provide shelter for greenhouses

sun through the glass and to trap this heat for as long as possible. The light and heat transmitting qualities of glass do however work two ways. The more glass there is in the greenhouse the more light and heat you let in while the sun is shining, but the more rapidly you loose trapped heat as soon as the sun stops shining. This is also true of any form of artificial heat used in the greenhouse and this applies particularly to all-glass greenhouses.

By contrast, a greenhouse which has solidly constructed brick, wood or metal base walls and therefore a smaller glass area, will admit less light and heat from the sun but the heat will be trapped for longer because of the better insulating properties of the solid base walls. Such greenhouses also cost less to heat artificially because of the smaller heat-losing glass area.

Most people nowadays prefer the modern glass-to-ground greenhouses because of their lower initial cost and easy erection. While these are for letting in the sun (which suits the tomatoes), they do suffer the disadvantage of being more costly to heat. This is obviously more important in colder or more exposed parts of the country than in milder areas. On balance, however, glass-to-ground greenhouses are best for tomato growing because they let in the maximum amount of light.

Any greenhouse used for tomatoes should be as draught-free as possible. It should also be free from drips. Ventilation must be good otherwise the tomatoes will 'cook' in the summer. Excessively high temperatures have just as detrimental an effect on the quality of the fruits as low temperatures. They can also damage the leaves and flowers which again will spoil the quality of the fruit. In addition they greatly increase the amount of water the plants will need. There is much to be said for having additional vents in roof *and* sides to prevent the greenhouse overheating.

Since tomatoes require a lot of water, the greenhouse must be equipped with a good water supply.

Where electric or gas heating is contemplated, the appropriate services should be conveniently available.

A final point about greenhouses is their height and strength. To do justice to tomatoes, which are a tall crop if well trained, the taller the greenhouse the better for big, full-season crops. For late crops, the height of the greenhouse is not so important as crops will be smaller and the plants will not grow so tall.

All types of greenhouses must be strong enough to be capable of giving support to healthy tomato plants which, when fully laden, can be very heavy.

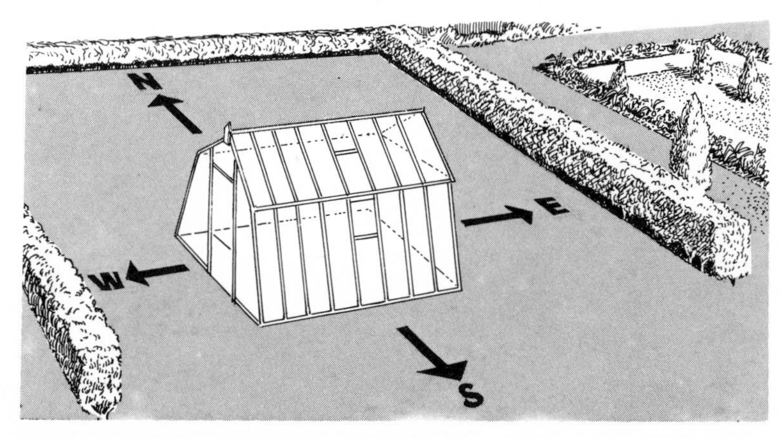

An ideal situation for a greenhouse. Note the use of hedges to provide shelter from the wind

POLYTHENE STRUCTURES

What has been said about greenhouses also applies to polythene greenhouses. Polythene, however, does not trap the sun's heat as well as glass, and polythene structures are therefore cooler at night. Condensation can also be a problem in polythene structures, especially in northern areas. The condensation can result in fruit being badly marked or spotted. Provision of support for the plants can also be a problem in polythene structures.

There are many types and forms of polythene structures. These range from those made of steel hoops, to 'do-it-yourself' structures made with wood. Amazingly good results can be achieved even with very crude structures made of polythene.

Even minimal protection of tomato plants can make all the difference between success and failure. By providing just that little bit of cover and extra heat, it is possible to produce ripe tomato fruits even in the dullest and coolest summers.

FRAMES AND CLOCHES

Various types of frames and cloches can be used for tomato growing with excellent results. Generally some modification of the usual growing methods is necessary. Alternatively, one can select special dwarf-growing varieties. The modern type of elevated frames are especially suitable for tomato culture.

PORCHES, CONSERVATORIES AND WINDOW SILLS

Tomatoes can be grown successfully in porches, conservatories or even on window sills, though it is usually necessary to grow the plants in containers of one kind or another. These can range from large pots to window boxes or troughs. Only conservatories, porches, or windows that face south and west are really suitable.

OUT OF DOORS

Tomatoes can be grown in a variety of ways out of doors, in a sunny, sheltered, south-facing border, preferably under a wall or fence, or in pots, bolsters (which will be described

This is what tomato growing is all about: a well-grown truss of rounded fruits ripening beautifully

A modern paraffin-fired heater—ideal for small greenhouses and excellent for heating houses where tomatoes are growing (Courtesy Aladdin Industries Ltd.)

The flowers of tomato plants deserve a closer look than they usually get

An ideal situation for the outdoor cultivation of tomatoes

later) or containers. They also make interesting plants for the terrace or patio. Some shelter in exposed areas is essential. The reflected light and heat given off by a brick or stone wall will greatly speed up the growth of the plants and the ripening of the fruit, while the warmth given off by any occupied building means that plants grown against a house wall can be planted out about a fortnight earlier than in open ground and left nearly a month longer to ripen. Beware, however, of planting them so close to the house that overhanging eaves prevent rain reaching their roots. Several varieties of tomato have been specially bred for outdoor culture. In cold, exposed areas, however, outdoor culture can be very difficult.

IN THE HOME

Large propagating cases or terraria, both of which should have provision for both heating and lighting can be used anywhere in the home where space can be found. Obviously, the nearer a window, and preferably a south or west-facing window, the less artificial heat and light will be needed. These containers are really only suitable for growing dwarf varieties, though they can also be used for starting plants to grow on in the garden where one does not have a greenhouse. They do however mean that people in flats can grow at least a small number of their own tomatoes.

3

General Needs of Tomato Plants

All green plants need light, warmth, water and a supply of vital nutrients to grow successfully. The tomato is, however, rather more fussy than most and requires some extra coddling. Because tomatoes are such an important commercial crop, a lot of research has been done into their precise growing needs, and these are now well understood. This takes the guesswork out of growing tomatoes. You may well have neighbours who have tried growing tomatoes by guesswork, and they will tell you all sorts of depressing stories about all the things that can go wrong with tomatoes. Ignore them for the moment. The important thing is to understand what the tomatoes themselves need. Later we can look at the things that might go wrong.

LIGHT

The first requirement of tomatoes is light, bright light, and plenty of it at all times. This can be a problem in Britain during the duller winter months. There should, however, be no real problems from spring until autumn, provided care is taken in siting the greenhouse correctly and avoiding obvious shade.

WARMTH

This is the next most important ingredient in successful tomato growing. It is necessary to think of both day and night *air* temperatures, and the temperature of the *soil* or compost. On average 56–59 °F (13–15 °C) by night and around 65–66 °F (18–19 °C) by day is best, starting ventilating at just over 74 °F (23 °C). If you can keep to about 58–60 °F (14–16 °C) or as near as possible day *and* night during the early stages of growth, this is best of all. A big difference between day and night air temperatures has a considerable adverse effect on the way tomato plants grow. When daytime temperatures are unduly high 68 °F (20 °C) and night temperatures unduly low 50–55 °F (10–13 °C) the plants become unbalanced in growth. This is because the plant leaves manufacture food during the warm day but cannot use this up at night if the temperature is too low. Plants grown under these conditions usually become curly leaved and squat in growth. Fruit can also become cracked, slow to form and rather small. When night temperatures are high, above 60 °F (16 °C), the plant becomes lanky and weak and fruit may be small. Many gardeners will be relying mainly on the natural temperature prevailing out of doors. The effect of this will quickly be seen, especially with outdoor culture. Tomato plants kept cold both night and day, 50–55 °F (10–13 °C) will grow very slowly. They may in fact succumb to root rot, especially if the soil or compost in which they are growing is also cold. When the plants are kept very warm, i.e. well above 68 °F (20 °C) all the time, they will grow quickly and usually produce smaller, ripe fruits earlier. It is worth noting that such plants may not produce a greater total crop, however.

Under glass, with some mild heat, ideal growing temperatures are readily achieved in Britain and countries of similar latitude between April and September, and without any heat from May until August. Thermostatic control of heating and ventilation is highly desirable. There will obviously be variations in any greenhouse, porch, polythene structure or frame, according to the area in which you live, and whether it is in an exposed or sheltered position.

Out of doors, the ideal temperature range we have been discussing can be difficult to achieve consistently in Britain, making it difficult to achieve good crops out of doors year after year.

Plants can, however, be grown perfectly well with no

artificial heat in greenhouses, frames and cloches. They will simply be later in ripening and produce smaller quantities of fruit.

POINTS TO REMEMBER ABOUT WARMTH

Early Crops Good level of heat in greenhouse. Good light. Soil warming especially useful. Early cropping is more easily achieved in southern regions of Britain and should only be attempted by the real enthusiast.

Mid-season Crops Less heat is required and this is the crop likely to appeal to the majority of gardeners.

Late Crops Heat, although not essential, is still helpful especially at night to prevent low temperature and excessive condensation.

Water Vigorously growing tomato plants require a lot of water. A great deal of the water actually applied is lost through evaporation and drainage so the plants only get a proportion. The water requirement of a tomato plant varies considerably according to the age of the plant and the weather. Young plants, especially if in a sand/peat compost, need very little water, but they quickly step up their requirement as they develop. As little as ¼ pint of water every 24 hours, or as much as 3 gallons per week could be required for established plants *over* 3 feet tall. Taking an average of 2 gallons per plant per week from mid-April until mid-September, this amounts to some 44 gallons per plant. Note that this is for border-grown plants. Considerably more water may be necessary with other cultural systems. This applies especially to ring and straw-bale culture, discussed later.

The water can be applied with a hosepipe or watering can. More sophisticated methods are discussed in Chapter 9.

PLANT NUTRIENTS

All growing plants require mineral elements known as plant nutrients. In this respect, the tomato is no different from other plants. However, it is a very greedy feeder, and requires greater amounts of nutrients than most plants. This is because it grows very fast, especially under glass. In particular, it needs a lot of nutrients to produce the large quantities of fruit which one expects. The following are the main nutrients needed by tomatoes.

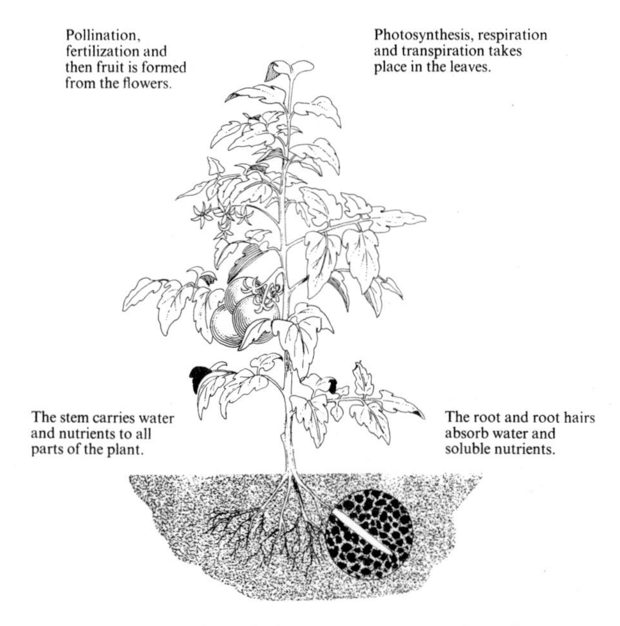

A diagrammatic representation of the way a tomato plant lives

Carbon Carbon dioxide gas is contained in the atmosphere. All green plants are dependent on carbon for the process of food manufacture which goes on in the leaves during daylight hours. There is usually sufficient carbon available in the air, especially if ventilation is carried out properly. Tomatoes can benefit greatly from extra supplies of carbon dioxide. The fumes given off by paraffin or natural gas heaters are sufficient to enrich the atmosphere.

Hydrogen and oxygen Two other important plant nutrients, are obtained through the water supplied to the plant and do not need supplementing.

SUPPLYING TOMATOES WITH THE OTHER MAJOR NUTRIENTS

The main chemical ingredients needed by tomato plants are nitrogen (for leaf and stem growth), phosphorus (for fruit and root development) and potash (for good fruit colour). It is possible to get extremely technical and even neurotic about the exact proportions, but this really is not necessary.

Broadly speaking, tomatoes need twice as much potash as nitrogen. They also require a smaller but definite amount of phosphorus. These needs are taken care of in proprietary tomato fertilizers, solid or liquid. The secret of growing tomatoes successfully is balancing the amount of nitrogen which gives soft growth to potash which gives harder growth. Once this has been appreciated, tomato growing becomes much simpler. Where tomatoes are being grown in borders, the soil should be of good quality and fairly rich. It always pays to apply some well-rotted farmyard manure, if this is available, at about 1 cwt per 6–7 square yards. Failing this, apply some peat at 9–10 lb per square yard. Residual nutrients in the soil, plus a general tomato base fertilizer applied at the rate of 8 oz per square yard, will meet about half the tomato's needs for the full season. A further 6 or 7 dressings, at the rate of 1 oz per square yard of tomato top dressing, applied every 10–14 days, or alternatively ½ oz per square yard applied every 5–7 days, will also be needed.

If you decide to use one of the many fertilizers which are diluted in water before use, make sure you always feed at the same frequency, i.e. every 10 days or every 14 days, not sometimes 10 days and sometimes 14 days. If you follow the dilution rate and recommendations which are issued with proprietary liquid fertilizers, you will not go far wrong.

Where systems other than border culture are involved, a fairly rich growing medium or compost should again be used at the outset. This rich compost should be about equal in food value to the border soil after it has been supplied with base fertilizer. Thereafter, solid or liquid fertilizers are used at the same frequency.

In addition to the nitrogen, potash and phosphorus content to the solid fertilizer or soluble liquid fertilizer you use, other elements are needed. Tomatoes, like many plants, require a certain amount of calcium. This is given as lime applied as ground limestone at about 8 oz per square yard to borders. Other elements such as Epsom salts (magnesium), plus others you may be less familiar with, are helpful. These are usually present in soil or already provided for in specially prepared proprietary composts. Many fertilizers, solid or liquid, contain a whole range of these elements.

Soil Analysis Now that you know the sort of soil tomatoes like, you need to find out whether the soil in your greenhouse is rich or otherwise in plant nutrients. It could, therefore, be worth sending your soil for analysis to your nearest Agricultural Institute or College. Alternatively, you can find out quite a lot with a soil-testing kit. The simplest form will tell you whether the soil has sufficient lime in it or not. All testing kits give full instructions and you should aim for what is called a pH figure of 6·5. For peat or peat/sand mixes, a pH figure of about 5·5–6 is satisfactory. Some testing kits also give information about nitrogen, phosphorus and potash. You do not really need to worry about these ingredients if you buy compost already mixed from a reliable source.

One word of warning however. Despite the prodigious appetite of the tomato, it is a plant which can be damaged by overfeeding, especially when young.

COMPOSTS FOR TOMATO GROWING

A bewildering range of composts can be bought at garden centres, garden sundries shops and nurseries. Some guidance on composts seems in order, and this is best considered in three sections.

1 Composts for Seed Sowing Tomato seeds contain sufficient food reserves to sustain the young plants for a short period. Where tomato seeds are to be germinated in a compost and then immediately moved on or pricked-off into pots or boxes, there need be very little in the way of food in the sowing medium. What is known as John Innes Seed Sowing Compost is ideal for seed sowing. So also are any one of the soilless seed sowing composts available.

It is not necessary to prick-off tomato seeds until they have reached a certain stage of development. They can be sown directly into pots or boxes where the plants can remain until they are planted out in their growing positions. Here a richer compost is necessary. Two courses of action can be followed. The first is to use a richer compost such as John Innes No 1 Potting Compost or soilless potting compost, which contains a higher level of nutrients. Alternatively, use the low nutrient seed composts but apply diluted liquid fertilizers from a very early stage. This is a very convenient arrangement with soilless composts. They usually have such physical characteristics that they are unable to store up reserve fertilizers unlike soil-containing composts.

2 Potting Composts We are concerned with the plants from the pricking-off stage until they are ready to be planted out. This involves a period which could be as long as 10–12 weeks or as little as 6–7 weeks or even less for later crops. The longer period in the pot will be occasioned by early sowing in poorer light. (The use of artificial light can shorten the propagation period.) John Innes No 1 or soilless potting compost is used for pricking-off purposes. Liquid fertilizers may still be required, especially in the soilless composts. They use up their food reserves more quickly than the soil containing John Innes Compost.

COMPOST TABLE FOR TOMATO GROWING

For Seed Sowing John Innes Seed Compost:

2 parts by bulk loam (sterilized)

1 part by bulk peat

1 part by bulk sand or gravel

To each bushel (22 × 10 × 10 inches) add 1½ oz super phosphate and ¾ oz ground limestone.

(N.B. a 2-gallon bucket equals ¼ bushel.)

Soilless compost—Any of the proprietary seed sowing or low-level nutrient composts.

Self-mixed compost—50% peat/50% sand (by bulk). To each bushel add 5–6 oz ground limestone and 2–3 oz Vitax Q4, Foremost CF1 or Osmocote.

For Potting John Innes Potting Compost No 1:

7 parts by bulk loam
3 parts by bulk peat
2 parts by bulk sand or gravel

To each bushel add $\frac{3}{4}$ oz ground limestone and 4 oz John Innes Base Fertilizer.

John Innes Base Fertilizer:

2 parts hoof and horn meal (by weight)
2 parts Superphosphates (by weight)
1 part Sulphate of Potash (by weight)

For cropping under ring cultures, in boxes etc, add

8 oz John Innes Base
$1\frac{1}{2}$ oz ground limestone

to make up the John Innes No 2 mixture.

Soilless Compost Any of the proprietary soilless growing composts. Self-mixed soilless compost:

50% peat/50% sand
or 75% peat/25% sand
or 100% peat
plus 3 oz ground limestone
3 oz magnesium lime
6 oz Vitax Q4
Foremost CF1 or Osmocote

It is important to mix composts accurately, and to warm them in the greenhouse for at least 48 hours before use.

4

Ways of Growing Tomatoes

There are a dozen or so different ways of growing tomatoes, and it is quite impossible to claim categorically that any one of them is infinitely superior to all the others. Each system has its merits and its disadvantages and ultimately one chooses the system that suits one's own needs best. For example, the gardener who erects a greenhouse on poor soil will usually find that some system avoiding the use of that soil is desirable. The converse of this is also true, particularly where the soil is of good quality and is known to be 'new' to tomatoes. All that can be done is outline the main options open to you, and leave the final choice to you.

BORDER CULTURE

In this method the existing or 'imported' soil of the greenhouse floor is used. The system has many merits where clean, new or sterilized soil is used. The main advantage is that less water is generally necessary than for other systems. This has great virtue for the gardener who is often away from home. Soil in bulk, however, is slow to warm up and can often be wet and inhospitable, especially at the beginning of the season. Plants may also grow too quickly and unproductively in rich soil, particularly in mild coastal areas.

One basic problem is that when the soil is used for tomatoes year after year, pests and diseases soon build up to a level where the plants fail unless the soil is effectively sterilized or

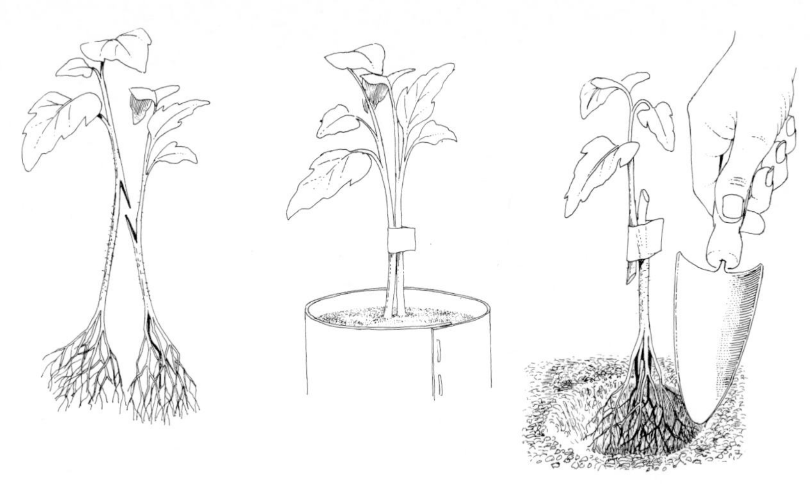

Grafting tomatoes by inarching technique

Removal of unwanted root stock from a grafted tomato plant

completely renewed. This is not without its problems. A frequent course of action is to use the border when the greenhouse is newly erected and then, when troubles start and plants die off or crop badly, turn to other methods.

GRAFTED PLANTS

This is a modification of border growing where, by a process of grafting, desirable cropping varieties are grafted on to root stocks which have resistance to most of the troubles encountered in older, unsterilized soils. The grafting process is, however, time-consuming. Grafted plants are tedious and expensive to produce and, if purchased from a nurseryman, cost at least twice as much as ordinary tomato plants on their own roots. Certain resistances to disease have now been introduced into ordinary tomato varieties. Grafting is nevertheless a highly successful cultural method with all the advantages, and few of the disadvantages, of growing in borders.

RING CULTURE

Under this system the plants are grown in specially treated paper (or whalehide) pots without bottoms in a limited quantity of clean compost (usually John Innes Potting No 2 renewed annually) spaced about 18–20 inches apart and placed on top of a 4–6-inch layer of clean, inert gravel, weathered ashes or sand. Constant watering and feeding are necessary. This means that a lot of time will be spent looking after the plants, especially early in the season before the roots develop extensively into the ashes. It is, however, a simple, relatively cheap system to set up. Providing care is taken with watering and feeding at the outset, it is possible to achieve very successful results. For this reason, ring culture suits many beginners, especially where soil in the greenhouse border is of doubtful quality and replacement of the soil is out of the question.

STRAW BALE CULTURE

Complete straw bales, or 9–10-inch sections of them (called wads), are induced to ferment or 'heat' by the application of water and a nitrogen-based fertilizer, almost like making compost in the garden. The bales are placed end to end on their flatter side or broken up into 9–10-inch wads. Both are laid on a layer of polythene which acts as a barrier to pests and diseases. Some planning will be necessary to accommodate the bales or wads to best advantage, as bales especially take up room.

Start treating the bales or wads about 2–3 weeks before the intended planting date by soaking them repeatedly with water. The greenhouse vents should be kept shut and sufficient heat put on, if necessary to give an air temperature of about 50 °F (10 °C). To each bale, nitrogen should be added in the form of Nitro Chalk or Nitram, at the rate of about 1½ lb per bale, well watered in. To find the quantity of Nitro Chalk or Nitram to apply to wads, divide 1½ lb by the number of wads obtained from each bale. For example, if you get six wads from each bale, apply ¼ lb per wad (1½ lb ÷ by 6).

After 3–4 days, apply another 1 lb per bale of the nitrogen again and water it in. Provided the bales are sufficiently wet and the temperature in the greenhouse is high enough, the

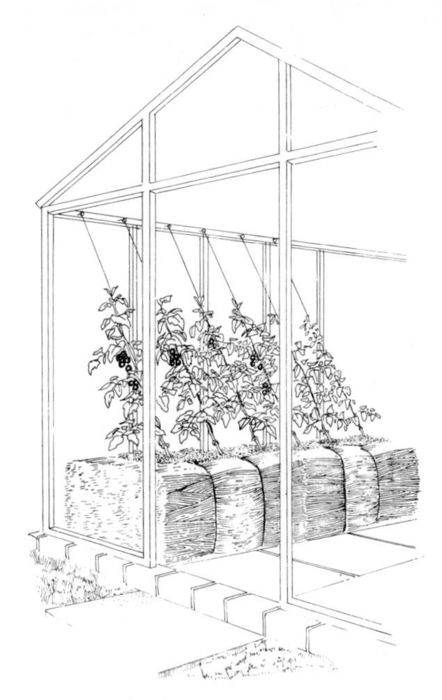

The straw bale method of tomato cultivation

bales should now be 'heating'. Finally apply $1\frac{1}{2}$ lb per bale of a good slow-acting fertilizer such as Vitax Q4 and still another $\frac{3}{4}$ lb of Nitro Chalk or Nitram and water these in. By now the bales should be piping hot, as you will feel if you put your hand into the centre. Wait until the temperature falls a bit, and then run a little 5 inch deep ridge of John Innes No 2 or soilless compost along the top of the bales. In a day or two, set the plants out with the roots merely covered by the compost at 3 per bale or 1 per wad. Alternatively, bales can be induced to heat by treating with Maxicrop Tomato Special which is a seaweed-based fertilizer.

As with ring culture, this is a method which suits the gardener with doubtful border soil. Apart from inducing the bales to ferment, watering and feeding can be tricky. This method has, however, proved very successful, even when used by complete novices, largely because the warmth of the bale gives the plants such an excellent start in life. In addition to the free heat, carbon dioxide gas is also generated, which further helps the plants to grow well.

CONTAINER AND HYDROPONIC SYSTEMS

This rather elevated description means quite simply that the tomatoes are grown in some form of container rather like any plant growing in a pot. The container need merely be large enough to keep the plants sufficiently well supplied with water and nutrients. As new soil or soilless compost is used every year (or perhaps two years), it is pointless and expensive to have containers which require a lot of compost.

Container growing could not be more simple. One merely selects a suitably sized container and fills this with suitable compost. The containers are spaced out at around 20–24 inches apart, and then when the compost is sufficiently warm the plants are inserted.

Growing tomatoes in polythene troughs

Suitable containers are 9-inch whalehide pots (with bottoms), polythene bags (12 × 12 × 12 inches) with drainage holes, holding about half a bushel of compost, or boxes which hold about 12 lb of John Innes No 2 compost. Boxes should be at least 12 inches deep. Treating the boxes with preservative makes them last a few years. Apple or orange boxes will do admirably for two plants. It is best to stand these boxes on benches or elevate them on bricks to give free drainage. Other containers include sausage-shaped 'bolsters' which are bought filled with peat compost with cut-out circles for the plants. Polythene buckets with three

1-inch drainage holes can be used with great success, especially with a vermiculite-based peat compost. Even large clay pots 9–10 inches in diameter can be used. This is an ideal system for porch or conservatory culture.

The choice of compost, whether soil-based or soilless, is not crucial but a matter of personal preference. It will be found, however, that plants are easier to water and feed in soil-based composts. The reason for this is that soil retains its nutrients for a longer period than soilless compost. With peat-based composts the nutrient supplies are much more quickly used up by the plant.

Peat trough systems involve the making up of polythene troughs 2 feet 9 inches to 3 feet wide and about 8 inches deep. While troughs can be provided with drainage, it is the experience of many gardeners and commercial growers that it is unnecessary to do so. All-peat compost is normally used for trough systems. While large quantities of peat compost are needed, the plants on the whole are easier to water and feed than in containers. Experience shows that the same compost can be used successfully for two or even three years without changing. Plants are usually set out about 12–14 inches apart in a double row 22 inches apart. Another system is to take out a trench in the greenhouse border 9 inches deep and 12 inches across into which polythene is laid. Peat compost is then put into the trench. Drainage holes seem unnecessary for a short-term crop.

PEAT MATTRESS SYSTEMS

Whalehide pots, 9-inch size (bottomless) filled with peat-based compost are set out in a 2-inch layer of peat in a shallow trough of polythene to which has been added 6–7 oz of ground limestone per bushel and 2 oz per square yard of a general fertilizer. As with ring culture, the plants initially contain their roots in the whalehide pot. They then send them out into the peat, which gives the plants additional supplies of water and nutrients.

WHALEHIDE POTS ON TOP OF SOIL

This is a very useful system where border soil is doubtful. Whalehide pots filled with John Innes No 2 compost are set

Ring culture showing the make-up of the bed

out 18–24 inches apart on top of the border soil, which is known to be diseased. For short-term crops this system is excellent. It must be appreciated that if there is a serious disease problem in the border soil this could eventually affect the plants in the pots. Alternatively, the pots can be placed on a layer of polythene or a shallow layer of clean soil or peat. This in effect resembles the peat mattress system.

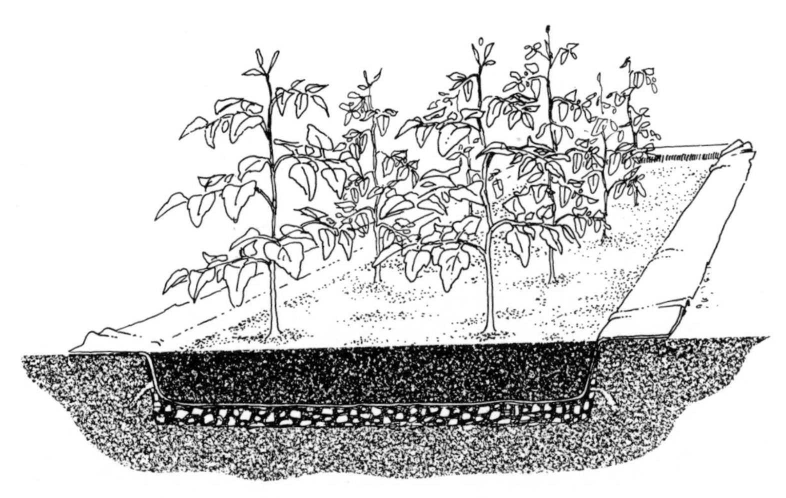

A simple hydroponics system

The peat bolster bale method of culture is one of the most trouble-free ways of growing tomatoes

An aluminium greenhouse like this one is perfect for tomato cultivation

SIMPLE HYDROPONIC SYSTEMS

This interesting and highly successful method involves growing the plants in a completely inert substance and feeding them with a rich liquid feed. First remove the soil to a depth of 8 inches and 3 feet wide, and place a 2-inch layer of gravel on the bottom of the trench, then line it with polythene. Fill this with sand, gravel or lignite (a form of brown coal). Drainage holes should be made in the side of the trench about 2 inches from the base. Provided complete liquid fertilizers such as Maxicrop Tomato Special are given from the outset, results can be really excellent.

More recently sawdust and shredded wood bark have been used for similar systems with good results.

SUMMARY OF CULTURAL METHODS

Border Growing Only for soil in greenhouse borders which is reasonably 'new' to tomatoes or has been sterilized. Varieties are now available which have certain resistances to soil-borne troubles.

Grafted Plants in Borders Excellent when borders are old and the soil cannot conveniently be sterilized or changed.

Ring Culture Use where the soil is 'played out' and diseased. Readily repeatable annually at low cost.

Straw Bales Use where soils are doubtful but bales take up too much room. Ideal when heat is in short supply.

Container Systems Can all give excellent results but constant attention to watering and feeding is necessary.

Simple Hydroponic Systems Very simple and give excellent results. Well worth a try by the adventurous gardener.

5

Planting Timetable and Plant Raising

The wild tomatoes in Peru that we talked about earlier let their fruits fall to the ground where in time they germinate and produce new tomato plants. Such a cycle of growth is common to most plants growing in the wild.

The tomatoes we grow in our greenhouses or out of doors would do the same if allowed. There are problems to this haphazard method of propagation. For one thing, one has little control when the plants start to grow. For another, it is not always satisfactory because a lot of the modern varieties (which will be discussed in Chapter 13) are F1 hybrids.

Tomato plants are usually raised from seed but they can be raised by rooting cuttings taken from sections of the stems. There are obvious difficulties here since one would need to keep tomato plants alive all winter to provide a source of cuttings. Nevertheless, cuttings from existing plants can be a very useful method of producing new plants for growing later in the season. Seed is, however, the most convenient and widely used method of producing new plants each year. Many people prefer simply to buy the plants from a nurseryman. This is an easy way of obtaining good young plants at reasonable cost.

Before discussing in detail either raising plants or buying them in, there is the question of when this should best be done to fit in with a convenient timetable. After all, once one knows

PLANT RAISING AND PLANTING TIMETABLE

Crop Definition	Seed Sowing	Potting	Spacing Out	Planting time	Picking fruit	Comments
Early. No Supplementary light	Mid Nov	Late Nov Early Dec	Mid Dec Mid Jan Late Jan (15–16 in)	1st–3rd week Feb	Mid Mar Early Apr	Only for the specialist gardener in good light area with plenty of heat.
Early. Supplementary light	Late Nov Early Dec	Early–late Dec give light for 17–21 days	On three occasions as necessary	Late Jan Early Feb	Mid Mar–Mid Apr	As above.
When raising plants in heated terraria or lit propagation cases the above programme applies. Plants for later planting can of course also be grown in this way.						
Second early	Early–Mid Dec	Late Dec–Early Jan	On three occasions as for early crop	Early Feb–Early Mar	Late Apr Early May	Still only a crop for the really well equipped gardener in a good light area.
Mid season	Late Dec–Early Jan	Mid Jan	On three occasions	Mid Mar–Early Apr	Early–Mid May	A crop for the gardener with a well-heated greenhouse.
Late	Mid Feb onwards	Late Feb	Twice as required	Mid–Late Apr	June	The ideal crop for the average gardener with moderate or no heating.
Later or cold	Early–Mid Mar until Apr	Late Mar–early Apr or later	Twice as required	Early May–June	July	A favourite crop for the beginner but growing season is greatly curtailed.

The above is only an approximate timetable which will obviously need to be adjusted slightly according to the district and facilities available.

how long a tomato plant takes from sowing the seed to the production of ripe fruit, then one only need know when one wants the fruit to ripen in order to be able to work back to when one needs to sow the seed or set the young plants in their permanent positions.

SOWING THE SEED

Tomatoes have small, light seeds. There are about 7000 to the ounce. A small packet will generally produce all the plants required for a small greenhouse. Pelleted seeds are also available and it is a simple matter to calculate how many actual pellets are required since they are bought by number. The process of germination of any seed, including that of the tomato, depends firstly on the softening of the hard coat which protects the seed. This allows the entry of moisture to start off the process of growth. It helps if the tomato seed can be soaked overnight in a saucer of water, although when wet they can be difficult to separate. With pelleted seed, the coating has first of all to be softened before the water reaches the seed contained in the pellet.

SOWING METHODS

For large quantities of seed, seed trays measuring 14×9×2 inches are used. For small quantities, pots or seed pans are ideal. A seed tray takes about 250–300 seeds, a 6-inch pot or pan about 60 seeds. Seed can also be sown individually in little peat or soil blocks. This is a lot easier with pelleted seed. Seed trays, pots or pans are filled loosely with seed sowing compost and stroked-off level. The compost is then pressed down with the bottom of another seed tray or pot so that the surface is level. The compost must be warmed by keeping it in a temperature of about 65 °F (18 °C) for a day or so. This can be readily achieved by using a propagating case or a light window sill with plastic trays and dome covers placed near radiators. Seed sowing receptacles can actually be placed on a radiator for a few hours.

To sow ordinary (not pelleted) seed, scatter evenly on the surface of the compost. It should then be covered over lightly by riddling some further compost through a fine riddle, before lightly firming and watering with plain water. Use only a fine-rose watering can from a low level, otherwise the seeds will rise to the surface, which is undesirable. The pots, pans or seed trays should then be covered with a sheet of polythene or glass, or a plastic dome cover. A covering of paper is also useful to avoid drying out until the seeds germinate. Pelleted seed can be sown on the *surface* of the compost, the distance apart depending on the receptacle. Generally 1–1½ inches is adequate unless sowing directly into containers such as Jiffy 7's. One or two seeds (pelleted or otherwise) per container are sown and, if both germinate, remove one. Seeds in containers should also be covered with paper and glass.

For germination to take place, a constant day and night air *and* soil temperature of 65 °F (18 °C) is required. Extremes of high and low temperatures and alternating high and low temperatures can play havoc with the seedlings, and may cause the resulting plants to be squat rogues or 'male' plants. These are relatively unproductive of fruit (although they may look healthy enough).

The seeds must be kept evenly moist, without extremes of becoming sodden or drying out, at an even temperature of 65 °F (18 °C), relatively dark and humid. The majority of the seedlings will appear between 8 and 12 days after sowing. All covering should then be removed.

Seedlings directly sown in large containers (i.e. pots of 3¼–4¼-inch size) are given the same treatment as other seedlings, although they obviously will not need potting-on.

POTTING

Two courses of action can now be followed. The first is to have pots of clay, plastic, peat or paper of approximately 4-inch size ready filled loosely with *warmed* compost. Into the centre of each a deep hole is made with finger or dibber. The seedling, handled by the seed leaves, is removed from its germination quarters and put into the hole and the compost firmed lightly round it with the fingers. The second is to half fill the pots with compost, then place the seedling (again held by the seed leaves) into the centre of the pot and graduallly ease in the compost. In both cases give the pot a sharp tap on the potting bench to firm up the compost and leave it ½ inch from the top rim of the pot. Follow this with a light watering. Directly sown seedlings in Jiffy 7's or small pots or blocks are potted by half filling the pot, popping in the small container, and gradually filling up with compost. Finally give a light watering. Seeds sown directly in larger pots are left where they are. Seedlings can also be potted into seed trays at 12 per tray.

Gardeners who wish to graft their own plants will find full details in specialized books (*The Complete Book of the Greenhouse*—by I. G. Walls, Ward Lock, 1974). Grafted plants are generally best bought from nurserymen.

TREATMENT OF YOUNG PLANTS

Good light, an even temperature, and regular but not over-watering are essential to keep young plants in prime condition. They must be kept growing well to avoid falling prey to fungal troubles. Take particular care to avoid over-watering in peat or plastic pots. Plants will relish one of the soluble feeds such as Maxicrop, Phostrogen or Bio. Plants also readily respond to foliar feeding at this time. Particular care must be taken to ensure regular feeding of plants or they will quickly starve.

Supplementary light can be given with fluorescent tubes or in a heated terraria in a poor light area. It is usual to give from 17 to 21 days of light treatment, allowing plants not more than 17 hours light in each 24 (7 hours of darkness). *This is important*. Light treatment makes all the difference in dull weather.

AIR TEMPERATURES

Try as far as possible to now keep night temperatures at about 58–60 °F (14–16 °C) and day temperatures at about 65 °F (18 °C). If it is too warm at night, plants will become leggy. If it is too cold, they will be bushy and squat. It is worthwhile investing in a maximum/minimum thermometer.

As the plants develop, they must be spaced out to get maximum benefit of light. Always make sure that the plants do not crowd each other out.

The most difficult period in which to propagate tomatoes is during December, January and February, due to the poor natural light. In March the light quickly improves and the plants grow quickly and are generally a lot easier to handle.

Tomato plants are usually available for purchase from nurseries from April until June. In many cases it is simpler to buy the plants as and when required, although growing tomatoes from seed is not really difficult and can be fascinating.

Plants are ready for planting when they are about 12–16 inches tall, including the pot. They should be of sturdy, balanced growth, with good leaf colour. Try to avoid spindly, soft plants which are easily toppled over.

As the season progresses and the weather out of doors gets warmer, propagation can be carried out at light window sills, in porches or in cloches or frames. Ideally, however, the best tomato plants are those raised in heated greenhouses and kept at a steady 60–65 °F (16–18 °C) throughout the propagating period.

6

Starting the Crop

GROWING IN BORDERS

Growing tomatoes in the 'floor' of a greenhouse is still the most popular method of culture. This is especially so when the soil is new to tomatoes or previous crops have given little trouble. The soil must be of good quality and of sufficient depth, with efficient drainage. Beware of wet, cold borders with rain seeping in from outside. If the soil is doubtful because of previous disease trouble with tomato crops it will require renewal or sterilization. If it has to be renewed, fresh soil which has not grown tomatoes before (or for many years) should be used.

Remember once again that it can be well worth paying the £2 or so for soil analysis where large areas are involved, or at least check the lime content with a pH kit (available at most garden shops and garden centres). Tomatoes like well-cultivated, rich soil with a pH of about 6·5 (which is slightly on the acid side).

First dig the soil well, removing any obnoxious weeds and, as discussed earlier, work in some well-rotted farmyard manure at the rate of about a barrowload per 6–8 square yards of border. Alternatively use peat at about 10–12 lb per square yard (there are 6–8 buckets of peat in a cwt which means a bucket contains about 16 lb). Add 2–3 oz of *extra* ground limestone for each bucket of peat. In round figures this is about 10–12 oz ground limestone per square yard. Well-made garden compost is also excellent.

After digging the soil, generally about 4–6 weeks before planting is intended, leave it rough until after watering as described below, which should be carried out 3–4 weeks before planting. Where soil is 'new' or re-soiling has taken place, do not firm or rake it yet.

PRE-PLANTING WATERING OR FLOODING

Soil should be fairly moist at planting time. Heat-sterilized soil should be watered really well, using a fine rose on the end of a hosepipe not only to moisten it but to flush out harmful chemicals. When soil has been chemically sterilized with liquids, watering may not be necessary. The only safe way to assess whether soil requires heavy watering or not is to have it professionally analysed. If this is not possible, it is always better to err a little on the side of safety by watering well to flush out the toxic remains of last year's fertilizers. Failure to ensure that the soil is moist enough at planting time and is not overloaded with fertilizer reserves could give rise to many problems. There are seldom any problems with new soils, which merely need to be moist.

After watering, allow sufficient time for the soil to dry out on the surface before raking it fairly level, if necessary firming up with the feet. Make sure that lime is applied (if necessary) before watering, so that it is well worked into the soil and will not conflict with any base fertilizer applied.

BASE FERTILIZER APPLICATIONS

In addition to farmyard manure, peat or compost and lime, 2–3 weeks after flooding out and a week or less before planting, it is best to apply a tomato base dressing. This will sustain the tomato plants with vital nutrients in the early stages.

Various types of fertilizers can be used as a tomato base. The main difference is the amount of potash they contain. Fertilizer manufacturers state the analysis on the bags or tins.

High Potash Bases are generally of around the following analysis:

Nitrogen 9% Phosphorus 9% Potash 17%

Medium or Standard Potash bases are:

Nitrogen 9%	Phosphorus 9%	Potash 13%

The John Innes Base Fertilizer is:

Nitrogen 5%	Phosphorus 7%	Potash 10%

which makes it suitable as a high potash base.

For early planting (February–March) it is always best to use a high potash base fertilizer at about 6 oz per square yard. When soil has been heat-sterilized also apply sulphate of potash at 2 oz per square yard. Where soil is new or has been chemically sterilized, *do not* apply the sulphate of potash.

For later tomatoes in new or unsterilized soil use Standard Potash base at 6 oz per square yard. Add 2 oz per square yard of sulphate of potash **where it has been heat-sterilized**.

It is useful to also apply Epsom salts (magnesium sulphate) at 2 oz per square yard as a source of magnesium, especially on older soils.

When soils are analysed professionally, follow the recommendations given regarding base application.

Carefully and evenly scatter base fertilizer 7 days or so before planting, finally raking the surface of the soil ready for planting when the soil is warm enough.

RING CULTURE PREPARATIONS

When, for various reasons, it has been decided to adopt the ring culture system, the first thing to do is remove the top 6 inches or so of soil.

AGGREGATE

A layer of weathered ashes, inert clean gravel, pebbles or very coarse sand, 6 inches in depth, should be evenly spread over the soil. Where there is real concern about pests or diseases in the underlying soil do not take any risks but put down a layer of polythene to seal it off. Do be careful, however, not to restrict drainage in any way and leave a moisture outlet.

Once the aggregate is laid, 9-inch whalehide pots (preferably without a base), which you can buy at any garden centre, should be placed out at 18–24 inches apart each way. Rings of linoleum, clay pots with enlarged drainage holes, boxes 10 × 10 inches, or old 2-gallon buckets with the bottoms cut out, can serve as rings. John Innes No 2 potting compost (*not* No 3 as is frequently recommended) should be placed in the rings, about 12–14 lb being required for each 9-inch whale-hide ring to fill it to within 2 inches of the top. More compost is needed for larger containers. Planting takes place when the compost in the rings is sufficiently warm.

Some gardeners prefer to keep their rings when planted in a smaller warm greenhouse and place them on top of the aggregate later. This can save a lot of heat. Rings with bottoms will be necessary to allow this.

PREPARATIONS FOR OTHER SYSTEMS

Pre-planting preparations for grafted plants are the same as described for border culture, although the soil will seldom be sterilized.

Reference has been made in Chapter 4 to the treatment of straw bales or wads. The ridge of John Innes No 2 compost or soilless media quickly warms up on top of the bales, providing a very congenial home for the plants.

With container systems it is simply a question of filling the containers and placing them at the appropriate distance apart. This is especially true of 'bolsters'.

Peat troughs are fairly simple to make up with wooden stakes, wires and staples, the polythene being lapped over the wire with a staple gun.

Peat trenches are simply laid into the ground and peat mattresses are a combination of a small shallow peat trough (2 inches deep) and whalehide pots filled with soilless compost on top of this. With whalehide pots on top of the existing soil the procedure is quite straightforward.

A brief description has been given regarding simple hydroponic systems.

Experience with any particular cultural system is very soon gained, the main consideration being correct spacing and adequate warmth at the outset.

7

Planting and Training

Despite the difference in setting up the various cultural systems described, the way in which the plants are set out and subsequently treated follows a basically similar pattern.

PLANTING DENSITY

Reference has already been made to planting distances. This is a subject of some controversy. Most greenhouses vary in size from 6×4 feet to 12×10 feet. Some may, of course, be larger. The plants should be so spaced as to obtain maximum light and growing height for each plant consistent with growing as many plants as will do well in the area available.

For an 8×6-foot greenhouse with the ridge running east/west, it is best to allow plants more space towards the south to let in the maximum light. When the greenhouse has the ridge running north/south the opposite is true. It does not automatically follow that the more plants in a greenhouse, the greater the total crop. The opposite is, in fact, often true since plants situated too close together crowd each other out and restrict entry of light.

Try to keep plants 18–22 inches apart, so that each plant has about $3\frac{1}{2}$–4 square feet. A good general rule is to divide the total ground area of the greenhouse (if it is all to be planted with tomatoes) by 4. Thus in an 8×6-foot greenhouse (48 square feet) one would put in 12 plants, planting 6 on each side of the path. It is a temptation to think in terms of 24 plants

(12 on each side), but you will find it almost impossible to do justice to 24 strongly growing plants. The results can, in fact, often be disastrous due to such overcrowding.

There must obviously be compromise according to the shape of the greenhouse, but it is worth bearing in mind these simple rules regarding spacing.

TEMPERATURE OF SOIL OR COMPOST

No matter which system of culture is undertaken, *never* plant until the soil or compost reaches a temperature of 56 °F (13 °C) at a depth of 4–6 inches. This should be carefully checked with a proper soil thermometer. Planting in cold soil or compost exposes the plants to serious disease risk.

Consult the timing programme given in Chapter 5 and plan things so that you put the tomato plants in their permanent places when the soil or compost temperature is adequate and when it is possible to hold *night* air temperatures at not less than 56 °F (13 °C) and *day* air temperatures in the region of 60 °F (16 °C). It has been explained earlier that timing depends on region, type of heating, and that variations in temperature are inevitable. Resulting growth and overall performance of the plants will depend largely on early treatment.

In general terms greenhouses located in the south, with mild heat, can be planted from about the first week of April onwards. Gardeners in the north, also using mild heat, should wait a week or so longer. Tomatoes can be planted in completely unheated greenhouses from about the end of April in the south, and from mid-May in the north. Seasons, however, can vary enormously and careful appraisal of the situations is the only real guide.

A final point to remember is that it is not merely a question of putting on the heating system overnight, prior to planting, and expecting soil or compost to be warm enough. This could take a week or more of heating. Soil or compost in containers will warm up much more quickly than border soil, especially in sunny weather.

IDEAL SIZE OF PLANTS

We referred earlier to the ideal stage for young plants for

planting. This is 12–16 inches or so in height, sturdy and of good colour. Such plants can usually be obtained from nurserymen from early April onwards at reasonable prices. Where grafted plants are available, these are generally about twice the price of ordinary plants. Do not buy plants from a nurseryman and leave them lying around in cold greenhouses or worse still in the boot of a car. Keep them in the light, at the right temperature (56 °F (13 °C) night and 60 °F (16 °C) day) and keep them well watered and cared for if for any reason planting has to be delayed.

PLANTING IN BORDERS (including grafted plants)

Assuming the border has been made ready for planting as described earlier, mark out the planting distances accurately with canes or sticks. Two courses of action are then open. The first is to take out planting holes 3–4 inches deep and the same across with a trowel. The plants can be stood in their pots on the surface beside the holes, which are left open to let the sun warm the soil. After a few days, the plants should be watered and carefully removed from the pots (unless in soil blocks or peat pots) and set out in contact with the now warm soil. Warm soil is then eased round the root-ball so that it is just covered. The seed leaves should be about 1 inch above the soil level. Deep planting should only be resorted to later in the year when the soil is really warm. Leggy, drawn plants can be laid on their side so that the height of the plant is kept reasonable.

The second way to plant really only applies to plants growing in soil blocks or peat pots. These are simply left on the surface of the soil and allowed to root downwards into the soil or growing media. While this helps to ensure that the plants are not checked by cold soil, there is the obvious danger of drying out, so they require constant attention to watering. This is not the method for the gardener who is away from home all day.

After planting, give the plants a light watering, about $\frac{1}{4}$ pint per plant, with plain water. If the soil is very dry more water should be given, but it is important to remember that over-watering the plants at this stage hinders growth rather than helps it.

The procedure for grafted plants is the same, although the variety root can be removed if thought desirable. This will give the plants a severe check to growth.

RING CULTURE

After placing the filled containers in position (18–22 inches or so apart) allow the John Innes compost to warm up before setting out the plants in the centre of the rings as described in border culture. Water them in carefully.

STRAW BALES AND WADS

There is usually little difficulty in getting the ridge of peat or compost suitably warm and plants (at three per bale) should merely have their root-ball laid in the compost and well watered in. Take great care to avoid drying out as there is always this risk on bales.

CONTAINER, TROUGH, BUCKET CULTURE, HYDROPONIC SYSTEMS ETC.

Here again warmth of the growing media is a prime consideration. There can be the attendant difficulty of watering plants in peat-based compost which either seems to get too wet or too dry. Peats, however, retain sufficient moisture for young plants without heavy watering.

ESTABLISHMENT OF PLANTS AND TRAINING

The tomato plant's main aim in life at this stage is to develop its roots into the growing medium. This is essential if it is to obtain the necessary food and water required for the growth of its stem, leaves, flowers and fruits. How well it does this will depend on the temperature of the growing medium (whether this is soil, peat compost or aggregate), *air temperature*, the amount of water available, and most important, the level of nutrients dissolved in the water available to the plant.

Plenty of warmth cannot compensate for lack of water and nutrients and vice versa.

There could be adequate warmth at the roots but coldness in the air. The converse could also be the case with disastrous results.

There are, however, certain basic rules that can be followed in order to reduce the risk of an unsuccessful tomato crop. Here are a few simple rules to follow. At the beginning only give sufficient water to prevent the plants wilting. Only by careful inspection can you decide just how much water to apply and how often. Watch the growing tip of the plant carefully. When it begins to freshen or turn a lighter colour this indicates that the plant roots are beginning to root into the growing medium and to develop.

Plants which 'sit still' are obviously not happy, being probably too cold, too wet or too dry. If there is sufficient warmth in the growing medium and in the air, and enough moisture at the roots, plants will start to grow almost immediately. Within a week they will almost have doubled in size.

The plant which sits still and does not grow is easy prey for the many diseases to which tomatoes readily fall prey. Briefly, if the plant's roots do not grow and develop the cells are soon attacked by diseases, even in clean soil or compost, and this applies especially to root rots.

Ideally, plants will develop better if the greenhouse or structure is kept close and humid with minimum ventilation and no draughts in the early stages. Help can be given by frequently 'damping down' the atmosphere by spraying water about with a watering can or hosepipe. This will lessen the need for applying water to the plants themselves and should be continued until you see the plants begin to grow strongly. Temperatures—a minimum of 56 °F (13 °C) night and 60–65 °F (16–18 °C) day with ventilation at 72–73 °F (22–23 °C) should be the rule. Try to avoid high temperatures during the day and low night temperatures since these cause squat plants with curled leaves. On the other hand, low day, high night temperatures make the plant become leggy and affect fruit size.

Certain varieties of tomatoes, such as 'Eurocross BB', favour warmer temperatures overall while varieties such as 'Arasta' seem to be happier with cooler conditions. Some tomato varieties are more vigorous growers than others. 'Ailsa Craig' (which is no longer as widely grown as it used to be) is a lot more vigorous than 'Moneymaker'.

Pinching out side-shoots. This is done between the nails of finger and thumb

Ring culture of tomatoes. The young plants are shown here one week after planting

These modernistic structures are ideal for growing tomatoes in small gardens or allotments
(Courtesy Q-Cloche Ltd.)

WATERING

Plants in borders will require very little water for the first few weeks, especially if the border was well watered at planting time. Damping down of the whole greenhouse, including paths, is desirable for quite a period after planting.

Plants in rings must be watched carefully as the rings will dry out quickly in sunny weather. The same is true of plants in boxes, pots and other free-standing containers. Damping down should again be frequently carried out when practical. Obviously, this is not so easy with plants grown in porches or on window sills.

Straw Bales Constant vigilance is required and it will be necessary to apply water on a regular basis to the bales with a hosepipe to keep them moist.

Troughs and Bolsters Peat compost may look dry, when in fact it is quite moist, so water with great care to avoid soaking. Once the plants get under way, which they will do very quickly in moist peat, they will put out a mass of fine feather roots. More balanced watering will be necessary. In bolsters, and polythene bags particularly, drying out can occur very suddenly on hot days. Watering can in fact be such a problem in bolsters and bags that only a trickle system of irrigation is likely to be entirely satisfactory.

SUPPORT FOR PLANTS

Plants must be supported before they start toppling, something which they will tend to do as they develop and with heavy damping down. Their natural habit is, after all, simply to sprawl. The use of tall canes, one per plant, while perfectly practical for porch or conservatory growing, has long since become outdated in the greenhouse, and has been replaced by wire and string methods.

Strong wire, securely fixed to the gables of the house as high as possible above the rows of plants is normally used. Some modification to the greenhouse may be necessary to achieve this. Next polypropylene twine or Fillis is looped around

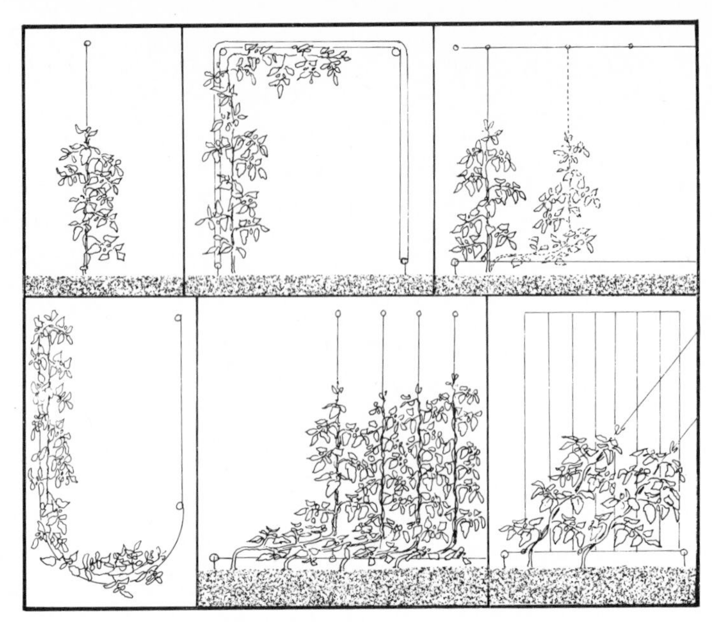

A sequence showing different systems of training tomatoes

the base of the plants below the first leaves with a non-slip knot and led up to the wires. Here it is tied with a bow knot, so that it can readily be undone for a reason which will be discussed later. As the plants grow, they are carefully twisted around the twine, consistently clockwise or anti-clockwise.

When straw bales are used, 18 inch long wires of $\frac{3}{16}$th gauge with a hook at one end are pushed into the bales. The twine is led from the hooks in the wire to the plants.

For other systems, especially in troughs, a strong wire is put up 12–15 inches above ground level and led along the row of plants. This is particularly important where layering systems are adopted to keep the fruit trusses from trailing on the growing medium.

TRAINING AND PRUNING SYSTEMS

The simplest system is to let the plants continue growing until they reach the top of the cane or strings. The top of the plant is then pinched out. All side shoots are completely removed when young (1–2 inches long). This is usually most easily done first thing in the morning or in the evening. To remove them during the day when temperatures are high can result in squashing them. They will usually snap off cleanly, but a clean, sharp knife can be used if you prefer.

Whether plants are restricted to one main stem or not depends a great deal on circumstances. A healthy, well-fed tomato plant can quite readily sustain up to three main stems quite easily. While it is a good idea to allow extra stems to form to make up the gap occasioned by the loss of a plant, some gardeners always grow plants with two or three stems. The vital difference is that they set the plants farther apart in the first place to allow for this. This does mean, of course, that one needs fewer plants. Sometimes plants develop two equal-sized leading shoots and it really matters little which one is removed. It is always more prudent, however, to retain the one already bearing a flower truss.

The technique for removing side shoots

With grafted plants, suckers from the root stock can be a nuisance and should be removed with a knife.

The lower leaves of the plant should be removed as they begin to turn yellow and flag. Left on the plants they will prevent the free movement of air and can become a focus of diseases which could spread to the whole plant. Generally speaking there is no need to start removing leaves until the

The use of string for training tomatoes

plants are 4–5 feet high. Leaves are detached from the plant by a clean up and down movement when the plants are well supplied with moisture. If they are removed with a knife, cut them back to the main stem.

DIFFERENT METHODS OF TRAINING PLANTS

While it was usual until a few years ago simply to grow plants vertically and stop them at a height of 6 feet or so, modern varieties, with their vigorous growth, coupled with more knowledge, have encouraged tomato growers to keep the plants growing on. A lot will depend, of course, on whether plants are planted early enough to allow this or are grown merely as a short-term crop. Where plants are planted in February (which is common practice in commercial circles) they can, by a system of layering, be induced to give 24 trusses or more. Even gardeners planting in March or April will find it possible by layering to reap a very much larger crop than would be the case if the plants were trained straight up and stopped at 6 or 7 trusses.

There are several fairly elaborate training systems, but the following is the most useful in a small greenhouse: firstly take the plants up to the top of the strings, then detach the string from the top wire and allow the plants to run along just above the ground, before tying the string to the wire again. It helps to make this easy if longer than necessary strings are used in the first place, with the loose knot previously described. It is best to do this when the bottom truss of fruit is picked to avoid it trailing on the ground. The plants are turned around at the ends of the rows.

Plants can also be trained on an angle from the outset by means of vertical and oblique strings or on 8 inch mesh nylon netting. This is a system which does tend to restrict means of access to the crop and is not too practical in small greenhouses.

The simple training system of growing on wires or string straight up to the maximum height is best for short-term crops. For the gardener wishing to plant early and continue cropping well into the autumn, the layering system is definitely worth considering, detaching and dropping the plants when they reach the wire.

8

Temperature Control, Watering and Feeding

Three factors are of prime importance during the growing season. These are temperature, watering and feeding.

Once a healthy young tomato plant begins to grow, the rate at which it will develop will depend not only on the temperature of air and growing medium, but also on how well you look after it. A young tomato plant requires very little attention, but as it grows it becomes more and more demanding of both water and nutrients.

The plant should be kept at an even temperature of around 60 °F (16 °C) during the day, dropping back to about 55–56 °F (13 °C) at night and rising to 70–75 °F (21–24 °C) during sunny days, when ventilation should be freely given. The opening of ventilators or the use of extractor fans lets the hot air escape and fresh, cooler air is brought in. The overall effect is reduction in temperature and humidity. Fresh air also contains vital carbon which plants require for their growth.

Temperature control can be very difficult in the small greenhouse. Although most people recognize the need to heat their greenhouses during cold weather, few people realize that proper ventilation during hot weather is equally important. Undesirably high temperatures can kill plants just as easily as undesirably low temperatures. Few manufacturers provide adequate means of ventilation in their greenhouse designs. The installation of automatic expansion-type vents is usually adequate for cooling the greenhouse, though more sophisticated devices such as extractor fans have the merit of positively changing the air in the greenhouse. For heating, a thermostatically controlled fan or convector heater is ideal. It is remarkable how well greenhouse temperatures can be controlled by a combination of properly installed ventilators and heaters.

Problems are most likely to arise during periods of exceptionally cold spring nights alternating with warm, sunny days. One often finds temperatures dropping below or soaring above those required. This need not be disastrous, provided the plants have made a good start and are healthy. Temperature control can be very difficult, especially for the gardener away from home during the day, unless he has a fully automated heating and ventilation system. The aim should be to try to avoid extremes of temperature, leaving manually operated vents open rather than risking overheating.

As the weather gets warmer artificial heating can usually be dispensed with completely. In many northern areas, however, some heat at night is always an advantage, except in periods of very warm weather. This is particularly true in the early autumn months when, if there is no heat at all at night, a lot of condensation may form and create ideal conditions for the development of diseases and cause the fruit to become marked. Condensation is a particular problem in plastic structures.

WATERING

More problems arise over watering than anything else. Young tomato plants require very little watering until they start to grow strongly. Different cultural systems make greater or lesser demands on water. This becomes very noticeable as the season progresses. While plants in borders may only require watering once a week or so, plants in rings or on bales often require watering daily. In bolsters, too, it will be found that watering is an almost daily necessity once the plants reach a certain size.

The quantities of water required by established tomato plants over 2½–3 feet tall are obviously closely related to the weather. Hot, sunny days encourage the plants to use up far more water than dull, cloudy days. The following are the general requirements:

Weather pattern	Water requirements per plant per 24 hours
Very dull, clouded over all day	$\frac{1}{4}$–$\frac{1}{2}$ pint
Dull but less overcast	$\frac{1}{2}$–$\frac{3}{4}$ pint
Sunny intervals	$1\frac{1}{4}$–$1\frac{1}{2}$ pints
Sunny with only the very odd cloud	2–$2\frac{1}{4}$ pints
Hot sun all day	3–$3\frac{1}{4}$ pints

This shows the extreme variation of watering requirements —a variation exaggerated still further when related to different cultural systems.

It is rather perturbing to realize that the water requirements of one plant can range from as little as 2 pints to as much as 3 gallons of water a week according to the weather.

Plants which are short of water will wilt in hot sun and often show little decayed areas on the tips of the leaves, especially at the top of the plant. In addition, flowers frequently drop off. Plants that are over-watered yellow at the bottom leaves. Continual over-watering often causes the roots to rot.

It is obviously tremendously important to give your tomato plants exactly the right amount of water, neither too much nor too little. It is not always easy to know just how much water plants need, or how much you are giving them. The 'Humex' moisture meter is a useful little gadget which, if pushed into the soil or growing media, gives a fair guide as to whether water is needed. It is particularly useful for peat composts, which can be tricky to water. This will tell you whether the soil is dry, moist or wet. If water is needed, you will need some way of knowing how much water you are giving. The only way to do this is to measure. You should know the capacity of your watering can, and you can work out how much water your hose delivers when turned on fully by timing how long it takes to fill a 2-gallon can.

SETTING THE FLOWERS AND DAMPING DOWN

A vital aspect of water application is the effect this has on the successful 'setting' of the flowers. This process is essential if fruits are to set and form properly. The need for damping down to give a humid atmosphere has been referred to for helping the plants to establish themselves. It will also greatly assist the setting of the fruits if the vents are shut down for an hour or so on sunny days and the plants are sprayed overhead. This practice can be continued until the plants are 3–4 feet tall.

Hormone preparations can be used to assist with the setting of flowers, which can be difficult in poor light areas or when the plants are over-vigorous.

Once plants are fully grown they tend to make the atmosphere in the greenhouse moister in any case, so no further aids to humidity are needed.

When growing plants in porches or on window sills damping down may not be possible. However, it is possible to spray the plants themselves.

It is not a good idea to water or spray plants last thing at night since this leaves the foliage too damp which can result in disease or marking of the fruit. At night the plants are dormant and droplets of moisture simply lie on the leaves and fruit. During daylight, when the plants are actively growing, such droplets are absorbed by the leaves and evaporate off the fruits.

FEEDING TOMATOES WITH LIQUID FEEDS

It would be nice to be able to say that feeding tomato plants was a simple, straightforward matter. In soil-based composts it certainly can be, but it tends to be rather more complicated in soilless composts.

Plants Growing in Soil Plants growing in well-prepared border soil (this includes grafted plants and plants growing in big boxes of soil) will usually be quite happy until the tomatoes on the first truss are well formed and about the size of marbles. This is the stage of development at which to start using a feed such as Phostrogen, Sangral, Maxicrop Tomato Special, or any of the proprietary liquid feeds. Try to obtain a liquid feed which is specially recommended for tomatoes, as it will contain the right balance of food for the tomato plants, usually the ratio of 2 parts potash to 1 part nitrogen.

These liquid feeds, diluted according to instructions, should be applied every 7–10 days to plants already watered with plain water.

You can, of course, mix your own liquid feed if you want to, suspending a bag of manure in a barrel of rainwater.

Plants in Soilless Composts Plants growing in soil have greater reserves of foods available to them than plants growing in soilless composts. Even if they are not fed for a week or two, nothing disastrous usually happens. This may not be the case in soilless composts. With these, it is better to start feeding *before* the fruit is formed on the first truss and to feed at twice the frequency as you would for plants growing in soil, i.e. every 4–5 days, using a balanced tomato liquid feed (ratio: 2 parts potash to 1 part nitrogen).

GENERAL RULES REGARDING FEEDING

Provided plants grow steadily and have a well balanced and even growth, there is no need to depart from this regular feeding programme. Two things can happen which may demand changes in this regime. Plants which become squat and hard and seem unable to grow vigorously are usually suffering from a lack of nitrogen. Sulphate of ammonia or dried blood, applied at the rate of 1½ oz in a gallon of water, should be watered carefully around the plants. Repeat the dose in a few days if no immediate response is apparent. This is generally given in addition to the normal feeding, but not simultaneously.

Should the plants grow too vigorously, producing masses of flowers on long stems, this generally means that they are getting too much nitrogen. It generally helps if sulphate of potash, diluted at 1 oz in a gallon of water (preferably warm, to help it dissolve), is given in addition to the normal standard liquid feed, but again not at the same time. If results are not apparent, repeat the dose until the plants are more restrained in growth.

Plants can often become over-vigorous if the days are hot and the nights are cold. Maintaining a more constant temperature, in addition to using sulphate of potash, usually solves the problem.

Solid Feeding Some gardeners prefer solid or granular fertilizers instead of liquid feeds. In this case tomato top dressings can be bought which are used according to directions. Do take care not to scatter the fertilizer too near the stem of the plant. Where extra nitrogen is needed, sulphate of ammonia or dried blood can also be used dry at about $\frac{1}{4}$–$\frac{1}{2}$ oz per plant, where too much is present, sulphate of potash at $\frac{1}{4}$ oz per plant. In all cases, flush in the fertilizer carefully with a light watering.

Overfeeding When feeding tomato plants with either liquid or solid fertilizers, it is better to feed more frequently with well-diluted or light feeds rather than infrequently with heavy applications.

Overfeeding clearly reveals itself by blackening of the leaf tips. When this happens, soak the borders or compost with plain water to flush out excess nutrients and start normal feeding again. Ironically, overfeeding can happen quite easily, when plants literally sit still. What happens is that the plants are in fact deprived of plant foods because the soil is so full of concentrated nutrients that the plant cannot use them.

Try at all times to feed *regularly* since missed feeds result in blotched, unevenly ripened fruits.

Overfeeding is a frequent cause of the malady known as *black* bottoms, especially in soilless composts.

9
Additional Equipment

Apart from the greenhouse, porch or polythene structure where you intend to grow your tomatoes, there are many extra items of equipment which can help to make things easier. All of these are available from well-stocked garden centres or specialist suppliers.

VENTILATION

The simplest self-operating systems of ventilation are those which use a simple expansion lift device fitted on to the ordinary manual vents. These expansion devices open and close the vents at pre-set temperatures. These systems are so simple and so reliable that they can hardly be faulted. If they do have a fault it is that they do not open ventilators sufficiently wide.

Extractor fans are also highly efficient. It is extremely important that they are used with air inlets, otherwise the plants may suffer from lack of oxygen. The fans are activated by a thermostat which switches it on and off at a pre-set temperature.

WATERING

A watering can is satisfactory for a small number of plants. The watering can may either be filled from a tap or from a reservoir such as a plastic tank. A hosepipe is, however, more practical, especially when several plants are grown.

Have the supply pipe as near as possible to the greenhouse with a screw connection. A good pressure of water helps. A vital piece of equipment is a rose to fit on the end of the hosepipe, as a forceful jet of water can damage the plants and wash the soil away from the roots. The rose is also necessary for damping down or spraying the plants to make the atmosphere more moist.

A sprayline is a very useful item of equipment. This allows the tomatoes to be watered overhead. It is also excellent for damping down the atmosphere. Control is usually manual on a small scale, with the spray lines connected to the tap. Alternatively, they can be operated by a hydrostat, a device which measures the amount of moisture in the soil or atmosphere and switches the sprayline on once the moisture content drops below a predetermined level. Trickle lines can also be connected up to a water supply for drip watering plants in bags, troughs or bolsters of peat. These can be turned on and off quickly and easily.

A dilutor is also a very useful piece of equipment since it will automatically apply diluted liquid feeding. It cannot, however, be connected directly up to the main water supply due to varying pressures. It can only be used in conjunction with a reservoir tank. It could be connected to the attic cistern in the house, or, alternatively a separate supply tank could be installed in the greenhouse on a ball-cock control.

Capillary watering systems on sand or fibreglass mats with header bottles are useful during the propagating stage.

LIGHTING

Fluorescent tube lights are a very good investment for keen tomato growers, and are especially useful during the propagation stage in dull winter weather. Three or four tubes, 2–3 inches apart, backed by a white reflector board, would be necessary to give a sufficient intensity of light over a reasonable area. Mercury fluorescent lamps specially designed for horticultural purposes are a viable alternative. They are highly efficient and will treat a large number of plants. It cannot be stressed strongly enough that all wiring for lights (and other electrical appliances) must be very carefully carried out with waterproofed fittings.

BENCH WARMING AND PROPAGATING CASES

A soil-warming cable with a loading of 10–12 watts per square foot, installed in a bench, is a tremendous help when raising tomato plants. Such cables are readily available from specialist suppliers and should be installed precisely according to the directions given with it. Even better are propagating cases with soil- and air-warming cables and, if necessary, lighting. Such equipment can be put to good use for propagating tomatoes and, later, for growing dwarf types.

POLYTHENE LINING OR DOUBLE GLAZING

Polythene lining is a good investment for tomato growing, especially early in the year when heat insulation is most needed. Thin, gauge 38 mu., clear polythene should be attached to the greenhouse, leaving the vents unobstructed.

SHADING EQUIPMENT

There is little need for shading in tomato culture, either by blinds or other means. Shading with much diluted emulsion paint or proprietary shading material on the outside of the greenhouse may be necessary when wilt diseases or eelworms attack the plants.

CARBON DIOXIDE EQUIPMENT

Sophisticated systems for enriching the atmosphere with carbon dioxide so that the plants will benefit from an extra source of carbon have been proved to be valuable for tomato plants in greenhouses. Unfortunately, there is no similar equipment available for general use. However, all the enrichment necessary is usually given off in the fumes of natural gas stoves or propane burners.

MOISTURE METERS

These provide a simple, accurate and quick way of checking whether soil-based or soilless composts are wet enough.

THERMOMETERS

A good maximum/minimum thermometer is essential for tomato growing to check that the correct range of temperatures is being maintained.

ASPIRATED SCREENS

An extremely useful item is an aspirated screen inside which are installed thermometers and thermostats controlling the greenhouse environment. An aspirated screen consists of an insulated box which shields instruments from the direct rays of the sun and from draughts, radiation loss, and other similar factors. A small continually running fan draws air over the instruments so that *true* and not distorted air temperatures are recorded.

HEATING GREENHOUSES

Greenhouses can be heated in a number of ways. It is important to ensure that the heating equipment is capable of heating the greenhouse to the level required. To do this, measure the total *surface* area of the greenhouse, multiply this by 1·4 and then by the temperature lift over outside temperatures required. For early tomatoes this could be as much as a 40 °F lift, 20 °F for mid-season tomatoes, and as little as 10 °F for the late season crop, merely to keep night temperatures at an acceptable level. The figure calculated is the number of BTUs (British Thermal Units) required from the heating appliance. A 6 × 8-foot greenhouse has an approximate heat loss of 300 BTUs so for a 10 °F lift it is 3000, a 20 °F lift 6000, a 40 °F lift 12,000 BTUs, and scaled down this is approximately 6, 12, and 24 BTUs per square foot of floor area. The output of all heating appliances is still stated in BTUs. For example a 2-kW convector heater is 2 × 3412 BTUs = approximately 7000 BTUs.

Oil heaters are normally in the 10,000–15,000 range but the cost of oil is now a major consideration. Gas heaters are in the 15,000–20,000 BTU range. Electric tubular heaters are usually rated at 60 W per foot. This means 17 feet of tubular heater for 1 kW. For most hot pipe systems where the water is heated by oil, coal, gas or electricity, estimate about 200 BTUs per foot of 4-inch pipe and 100 BTUs per foot for 1¼-inch pipe, which again allows calculation for the necessary length of pipe. All such information is obtainable from the manufacturers. It is important to find out *exactly* what the heating appliance is capable of achieving in relation to the size of the greenhouse, and the time of year it will be used. Perhaps one of the most useful heaters for tomato growing in the average greenhouse is the 2500 W electric fan heater with thermostatic control. In modern terms, however, the Shilton natural gas burner is proving very useful and economic.

'Tiny Tom', perfect for the small garden or for growing in porches or on windowsills

One of the modern yellow tomatoes 'Golden Dawn'. The fruits look unripe, but they are exceptionally tasty

A plastic structure, ideal for growing tomatoes in rural areas. The lettuces shown here make an ideal follow-on crop

10

Pests, Diseases and other Problems

Not so many years ago there were many troubles which were almost certain to attack the tomato plant in greenhouse or out of doors. Thanks to the genius of plant breeders and chemists many of these ills can now be successfully overcome. Tomato troubles are often straightforward and easy to diagnose. Conversely, plants may wilt or look generally unhappy and it is difficult to see why this should be. Nor in fact may one specific pest or disease be the culprit. Broadly speaking troubles come under three headings—*DISEASES, PESTS, PHYSIOLOGICAL DISORDERS*. If the picture sounds gloomy do remember that every plant that grows, tomato, chrysanthemum or shrub, has much the same sort of bombardment to contend with.

DISEASES

A great many of the diseases for tomatoes are caused by the minute spores or 'seed' fungus. These spores may either be air-borne or carried in the soil. Clean soil or growing media is an essential preventive measure. At any stage of the plants' growth they can attach themselves to the root, stem, leaves or flowers. Like a seed, these spores germinate and send their 'roots' into the portion of plant they have attacked. They then break down the plant cells for their own food, destroying the host plant in the process.

This process of destroying plant tissue can be disastrous, especially when some vital part of the plant is affected. When the roots of a plant are badly attacked they are simply not capable of drawing in water and nutrients. If the cells in the stem of the plant which conduct the water and nutrient from the root to the leaves are obstructed or damaged, serious wilting occurs. If fruit is attacked the fungus renders it inedible.

A few tomato diseases are not caused by fungal disorders, but by bacterial organisms. These act in much the same way. Control of these diseases lies either in not allowing them to attack the plant in the first instance, or in killing them off if they do attack the plant—if possible without damage to the plant.

Other forms of disease are known as plant viruses. They get into the sap of the plant to live there. They make a real nuisance of themselves to the great detriment of the plant.

PHYSIOLOGICAL TROUBLES (including nutrient upset)

Under this heading come all those troubles related to the effects of too low or too high temperatures, too much or too little water, too many or too few plant nutrients; plus several other circumstances. It is essential to realize that plants, including tomatoes, can have several things wrong with them simultaneously.

PESTS

Various insects like the taste of tomato plants and they make it their business to find them and feed off them. As with disease prevention, it is common sense to use pest-free soil or growing medium, but the insects are still attracted to plants and it is often necessary to use a chemical to deter them or kill them without damage to the tomato plant.

TROUBLES WITH YOUNG PLANTS

'Damping Off' When seed is sown the little plants sometimes fall over before they are large enough to be pricked-off. Close examination of the little plants shows that the stem at ground level is 'puckered' or shrivelled. The trouble is caused by several different fungal diseases. These diseases are often

Root rot of the neck or collar of the tomato plant — Damping off

present in dirty boxes, pots, water or soil. They also drift about in the air and merely attack the very soft tissue of the young plants.

Prevention and Control It pays not to allow the seedlings to get too hot or humid as this makes them very soft and disease prone. As soon as germination has taken place the seedlings should be kept slightly cooler and given more air. It can be fatal to keep the dome lids or sheets of polythene on seed trays after germination. Apart from using clean receptacles, soilless media is preferable to sterilized soil-based growing media (i.e. John Innes compost). The release of ammonia gas by these will invariably damage the sensitive little plants and leave the root open to attack by disease spores. Use a 50/50 peat/sand mix with *lime only* with the application of diluted liquid food once the seeds have germinated. This is now common practice. Seedlings generally, including tomatoes, are very prone to damage either by chilling or overwatering. So care must be taken in this direction.

The use of copper-based fungicides (such as Cheshunt Compound) is an excellent preventive measure. It is applied as soon as the seeds have germinated.

When the seedlings are potted they must be handled with great care to avoid physical damage.

Root Diseases—affecting both young and older plants There are a number of different species or forms of fungus that attack tomatoes causing damage to roots. While the keen gardener may want to know *why* his plants died and the name of the disease responsible, the actual variety of fungus is really only of academic interest. In practical terms it is much more important to avoid root rots in the first place. In a great many cases this can be done by using common sense.

It must be appreciated that root rots are almost inevitable if tomato plants are set out in a cold soil—i.e. one below 55–56 °F (13 °C), whether at the potting or planting stage. The root tissue of the tomato plant is not especially delicate, but it simply rots if it cannot grow, and it cannot grow in cold soils. Root rots are almost inevitable if young plants are put into cold growing media and all the chemicals and products on the market cannot do much about this. It is better to prevent trouble by avoiding cold soils (or soilless media) at the outset. Once a soil (or soilless) media has root rot disease spores in it, the disease can remain dormant for a long time, unless killed by sterilization. Use fresh, clean soil or soilless media for potting and planting, especially in the latter case if there is a history of root rots from the previous year.

Symptoms In addition to general lack of vigour and wilting, inspection of the roots can determine whether they are white and healthy, or brown and rotted. When the plants are removed at the end of the growing season the roots should always be examined for disease.

Root rots can take several different forms. Plants can be attacked at the actual collar or neck of the plant (called collar rot), at the extremity of the roots (called toe rot) and invariably caused by cold soil at lower depths, and by corky rot when the roots look clubbed—almost like club root of brassicas (the cabbage family).

Prevention and Cure With all root rots, prevention is better than cure. It is in fact doubtful if any real cure can be effected once plants are badly attacked. Undoubtedly 'new' soil or growing media which has not grown tomatoes before is the best preventive measure of all. All systems other than border culture lend themselves fairly readily to the use of new soil or growing media, if not every year, at least every second year (three years may be possible with peat compost in troughs).

Border soil can be sterilized by heat or chemicals but this is seldom convenient or practical. To renew border soil is laborious, always provided one can find reliable soil in the first place. Bear in mind that soils from vegetable gardens which have grown potatoes for many years are very likely to contain cysts of potato eelworm which can also attack tomatoes.

The use of grafted plants gets over most of the problems of older, unsterilized borders. The resistances which have been bred into the root stock do not however include all root rots. Nevertheless the sheer vigour of the root stock appears to do much to offset general trouble. This is the case not only with root rots but with other maladies which will be referred to later.

Where plants do fall prey to root rots, as indicated by general debility and wilting, it helps to mulch with a deep layer of peat (plus some lime). This, if built up round the necks of the plants will induce surface roots to form. This treatment may not always be entirely successful, though it may sustain the plant sufficiently well to produce a reasonable crop of fruit.

Some control may also be possible by the application of chemicals such as Sterizal or Dithane (watered in around the plant). The use of Benlate, which is a systemic fungicide absorbed by the plant, may also help.

Some of the new tomato varieties now have resistance to wilt diseases. It seems likely therefore that varieties with root rot resistance will in time be developed.

Without doubt, however, the best of all prevention for root rots is to avoid cold soil or growing medium from the time the tomatoes are planted. If the plants are home-raised, try to avoid checks and irregularities at the propagating stage.

Wilt Diseases (Verticillium and Fusarium) To most gardeners it little matters whether a plant has fusarium or verticillium wilt. The effect on the plant is to cause it to wilt badly during the day, though the plants seem to recover overnight when it is cooler. The bottom leaves can yellow badly, and only one side of the plant may be affected. Fusarium wilt is more prevalent in the warmer regions, verticillium in the cooler regions. Both diseases affect the water-conducting tissues at the base of the stem.

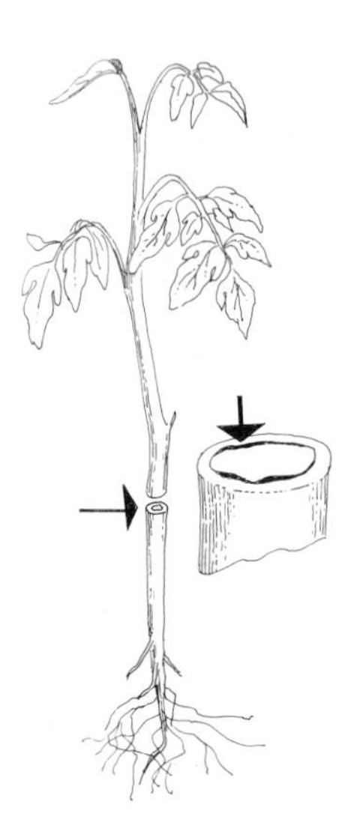

Wilt disease

Prevention These two diseases are soil-borne fungi. Root stock KNVF is resistant to both wilts (the V referring to the Verticillium, and the F to the Fusarium). Certain varieties such as 'Hollandbrid' and 'Eurobrid' also have inbred resistance. To avoid attack use clean or sterilized soil, or new soilless compost. Erratic temperatures should be avoided. Root rots are also a frequent occurrence along with wilts and the combined effect is doubly disastrous. Badly affected plants, if at the stage where they wilt badly all the time and do not recover at night, should be removed and burned.

Some means of control of verticillium may be effected as follows: first of all mulch the plants with peat or other material to a depth of 2 inches. Then shut the vents, put on heat and damp down with a hosepipe. Apply shading to outside of glass. The object is to 'cook' the plants for about 14 days at temperatures above 77 °F (25 °C). This is to induce new roots to form into the mulch from the stem above the diseased roots and so give the plants a new lease of life. This treatment is not very effective in bad cases. If fusarium disease is the problem this treatment will be completely ineffective.

The use of the product Benlate (a systemic fungicide which is freely available at garden centres and shops) has given reasonable results with both verticillium and fusarium wilt. It seems likely that other chemicals similar to Benlate and made specifically for wilt diseases, will become available in the near future.

Botrytis or Grey Mould Without doubt this common disease is the greatest bane to the tomato grower. Grey mould or botrytis is the disease which grows on any soft organic debris. Anything soft in the muggy atmosphere of a greenhouse or porch is prone to attack by grey mould. This is especially so in the late summer when the nights are damp. Damaged leaves, rotting lower leaves, scars where a leaf or side shoot was removed, petals of flowers and the fruit itself can all be attacked vigorously. 'Ghost spotting' occurs when the skin of the fruit is only partially attacked by quickly developing spores overnight. The spores dry up during the day, but the fruit is left badly marked. More seriously, if the grey mould attacks where the fruit is attached to the stem, the fruit will drop off before it is ripe.

Prevention One of the best preventive measures is to ensure that plants are not needlessly damaged. More important still, that the atmosphere is kept as dry as possible by good ventilation. Avoidance of watering at night is also important. Another point is to avoid planting too densely since this restricts air movement through the plants. Care should be taken not to overfeed the plants with nitrogen, which keeps them too soft.

Some heat at night, coupled with normal ventilation, can do much to avoid botrytis, especially in the cooler areas of the north. When on holiday it is better to leave ventilators fully open night and day than to risk the greenhouse fogging up overnight. To avoid ghost spotting of fruit, some ventilation at night, preferably with heat too, is advisable in colder regions after August. Some varieties are notably soft and prone to attack. 'Moneymaker' is typical and should not be grown in damp areas. In addition it is a sound practice to remove the lower leaves regularly to allow free movement of the air through the plants.

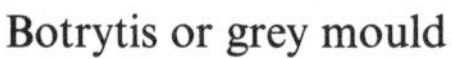

Botrytis or grey mould

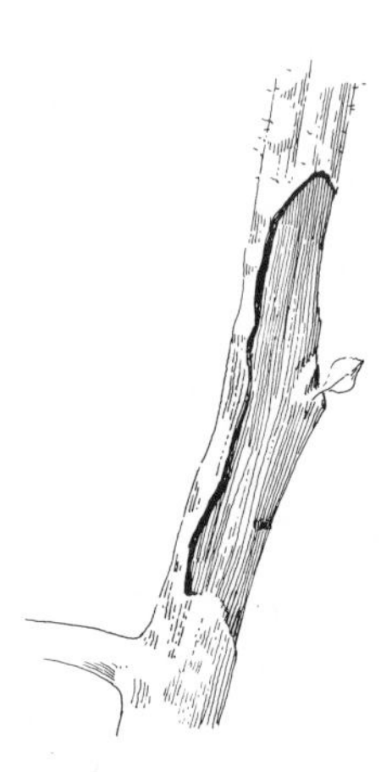

Stem rot

Control When superficial infection occurs (the symptoms of this are ugly black scars), scrape away the damaged area down to clean tissue. Then paint the wound with Benlate paste, flowers of sulphur or creosote.

Recent experiments have shown that a high degree of preventive success can be achieved by using Benlate before the plant is attacked. It can also be used after initial attack.

Once the disease penetrates into the vital conducting tissue of the plant it can cause the whole upper part to collapse. The plant should then be removed, otherwise there is a strong probability that it will infect other plants.

Didymella Stem Rot This disease, when it does occur, can cause tremendous damage. Fortunately it is not a common trouble. The stem is attacked, usually a few inches above ground level. Brown scars marked with slimy black dots are visible to the naked eye. Once this disease gets a hold, every plant in the greenhouse can be lost. If you even suspect the presence of this disease remove the suspect plant immediately and burn it. In addition spray the lower stems of all the remaining plants with a copper-based fungicide. At the end of the season remove and *burn* all plants. Then wash the greenhouse very carefully indeed with a bucket of detergent and a long-handled brush.

Leaf Mould Disease The symptoms of this disease are yellow patches on the top of the leaves. Underneath these patches (on the underside of leaf) the brown, felt-like clusters of the fungal growth can be seen. There are several strains of leaf mould disease, but what are called A and B strains are the most common in Britain. The disease is worst in humid areas and in greenhouses when there is excessive condensation overnight. Constant ventilation is the essential preventive measure, preferably coupled with some heat (as for botrytis control). In modern terms, leaf mould will not be a problem since most modern F1 varieties are resistant to the A and B

Leaf mould

strains. Most seed catalogues state which varieties are resistant to this disease. The well-known varieties 'Ailsa Craig' and 'Moneymaker' are NOT resistant, but there are many varieties bred from these two excellent varieties which are.

When non-resistant varieties are grown, or when the disease occurs due to a resistance breakdown, spray with the chemicals nabam and zineb. Careful removal of affected leaves also helps.

Phytophthora Stem Rot With this disease the very base of the stem is attacked by brown or black fungus growths which in time produce a white, fluffy growth. Careless fertilizer application is often the cause of the attack. Remove and burn affected plants if they die, and sterilize or renew soil for the next season.

Buck Eye Rot of the Fruit This disease is quickly recognized by marks on fruit almost like the eye of an owl. Infection is caused by soil-splash. Try to avoid careless watering and spray the lower parts of plants and fruit with a copper-based fungicide.

Potato Blight Mainly a disease of outdoor tomato crops: brown areas appear on leaves, and dark streaks on the stems. Fruit may also be attacked, rendering it useless. Spray regularly with a copper-based potato blight fungicide, following manufacturer's instructions.

VIRUS DISEASES

Many different virus diseases attack tomatoes, the chief culprit being tomato mosaic virus (TMV). This can check seedlings and cause leaf distortion. More important is the effect it can have on the ability of the plant to set fruit. More serious forms of virus cause the leaves to go fern-like and the plant to become seriously stunted.

There is little one can do to control the virus. The milder forms will check the plant for a period, after which the plant may recover and grow out of the symptoms to a certain extent. This can be helped by the application of a quick-acting form of nitrogen.

TOMATO PESTS

Eelworm There are two forms of eelworm. The most serious is the **Potato Eelworm**.

Infestation comes originally from land where potatoes have been repeatedly grown. It is often possible to have your soil professionally checked for eelworm. Affected plants will wilt in the hot sun, and will form yellow and dead patches on the lower leaves. The eelworm are microscopic creatures and nibble away the root hairs, restricting the uptake of moisture and dissolved nutrients. Plants attacked by eelworm do not necessarily die. The eelworm destroy the protective layer of the root, and this enables fungus diseases to attack the plant more easily. These are perhaps more damaging in the long term than the eelworm itself. A few new roots will usually form at the soil surface. Confirmation of potato eelworm attack can be found by carefully examining the root hairs. Yellow- or red-coloured 'cysts' the size of pin heads are seen. These contain the larvae of the eelworm and can remain in the soil for years and infect the subsequent crops of tomatoes or potatoes.

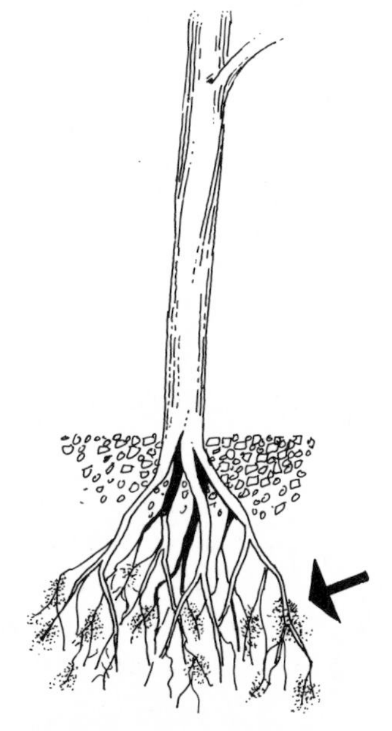

Eelworm infestation

Root knot eelworm do not infect tomatoes so frequently. They can, however, be very damaging, causing similar symptoms to those of potato eelworm but with more wilting. The tomato roots develop gall-like growths on them.

Control of both types lies in sterilization of the growing media, preferably by heat, or in complete replacement of the growing media with eelworm-free soil. Alternatively tomatoes can be grown on a different cultural system in clean growing media. Excellent control of root knot eelworm can be gained by using grafted root stock (KN or KNVF).

Aphids The symptoms are yellowing of leaves which become covered with sticky exudations. Spray young plants with Malathion. Use gamma BHC on older plants, but interchange the spray material regularly to avoid a build up of resistance.

Whitefly This can be very troublesome indeed and causes serious wilting of leaves and marking of fruit. The white flies rise in a cloud when the plants are disturbed. Spray with Malathion or Diazinon and again vary treatment.

Red Spider Mite This shows itself as spotted foliage, curled leaves and close examination will reveal minute spiders on underside of leaves. Use Azobenzene smoke and spray with Malathion.

Thrips The main trouble with this pest, which causes spotting of the foliage, is that it spreads virus infection. Spray with Malathion—two applications at 14-day intervals.

Springtails Young plants develop pinholes and signs of scraping on the leaves. Apply BHC or Malathion direct to soil.

Tomato Moth The leaves are stripped, and the fruit becomes holed. Spray with Dichlorvos.

Tomato Leafminer Normally only a problem in the south of England. The larvae tunnel into the seed leaves on younger plants. Spray the soil and seedlings with Diazinon.

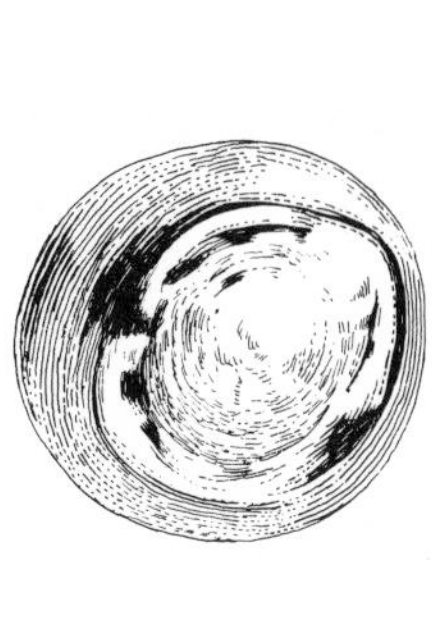

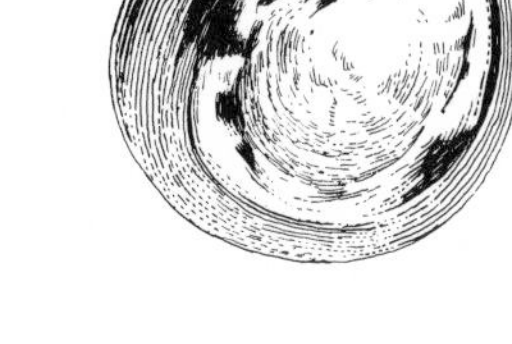

Blossom end rot

Dry set of the tomato flowers

Physiological Disorders A number of troubles occur which are not specifically related to a pest or disease, but are due to such things as variations in temperature levels and watering frequencies.

Blossom End Rot This condition reveals itself when the ends of fruits turn black. It is usually due to lack of water at a critical stage of growth. It can also be due to overfeeding (which results in plants being unable to absorb water). To aid recovery, water regularly and feed at correct rates.

Dry Set This occurs if, during the process of fertilization, so necessary for fruit formation, the pollen goes dry. Only pin-head fruits are formed. It is usually caused by dryness of the atmosphere in periods of high temperature, but can also be caused by the presence of infertile pollen which can be due to a variety of reasons. Virus disease may also be involved. Apart from levelling off the temperatures and damping down the atmosphere not a great deal can be done.

Flower Abortion The problem here is that, although flowers set, the fruit does not form properly, being small or 'chat' size. This is more of a problem with early crops when light intensities are low and plants are extremely vigorous in growth. It is important to raise plants under good conditions of light and heat. This is all that can be done.

Where flower trusses do not form properly This is due almost entirely to excess vigour in the plants, especially during the formation of the early trusses. The answer is to 'harden' growth by applying more potash and to keep plants on the dry side in the early stages of growth. High day/low night temperatures may also be a contributory factor.

Blotchy ripening

Dropping Flowers This is the complaint where the flowers drop off at the knuckles. It is due almost entirely to lack of sufficient water at the roots.

DISORDERS OF THE FRUIT

Blotchy Ripening This is caused by irregular watering and feeding, particularly with vigorous varieties. Where plants have been fed regularly with liquid or solid fertilizers and they are suddenly given an excess of plain water, a batch of blotchy

Curling of tomato leaves

fruit can occur. So try to avoid such irregularities. Very warm weather can also cause blotchy fruit.

Greenback This is where the shoulder or top of the fruit remains green and fails to ripen. It is more prevalent in 'Ailsa Craig', a greenback type, than in 'Moneymaker', a non-greenback type. Apart from variety characteristics, it is encouraged by lack of potash in the soil. Over-defoliation and high temperatures can also be contributory factors. Step up potash levels and shade glass in extreme cases.

Bronzing A dead layer of cells develops below the skin due either to excessively high temperatures, virus or lack of the element boron. It usually cures itself.

Splitting of Fruit This is due to varying weather and irregular watering. It also occurs when fruit has grown cold and has taken a long time to develop. Try to even out temperatures and be more regular with watering.

FOLIAGE TROUBLES

Curling of Leaves This problem is due almost entirely to big differences between day and night temperatures. Plants manufacture a surplus of food in their leaves during the day which they then cannot use up at night under cool conditions.

Oedema Bumps or blotches appear on stems, and to a lesser extent on the leaves. It is due to excessive humidity in the atmosphere. Give adequate ventilation and try to keep the atmosphere less humid.

Silvering Here sections of the leaves turn a light colour or go silvery. The cause is not yet known but it does not usually persist.

MINERAL DEFICIENCIES

Magnesium Deficiency Yellowing and lining between the leaf veins. Will gradually progress up the plant. Spray with 2% magnesium salts, ½ lb in 2½ gallons on several occasions. Reduce potash applications.

Iron Deficiency Yellowing and whitening of whole leaf. Younger leaves affected first. Avoid over-liming.

11

Tomatoes Out of Doors

Quite a lot has been said already about the warmth which tomato plants require if they are to produce ripe, succulent fruits. It is worth emphasizing once more that sun and warmth are really important, especially when considering the cultivation of tomatoes out of doors. Assuming the young plants have been raised under cover, planting time varies from May in the far south or Channel Islands to June in the far north. A good guide as to when to plant out of doors is when early potatoes have their leaves well through the ground. This is because potatoes and tomatoes are closely related. Both are, in the early stages of growth, very susceptible to cold weather.

Whether you can produce fully formed tomatoes completely out of doors will depend almost entirely on where you live. Broadly speaking, the farther south you live, the better. If, however, some protection can be provided by a polythene cover, polythene tunnel, cloche or frame, this can make all the difference. There is probably nowhere in Britain where tomatoes can be grown really well without some protection.

Remember that there are varieties specially bred for outdoor cultivation. These can ripen their fruit at lower temperatures than the indoor varieties and are smaller-growing in habit. Tomatoes also make interesting pot subjects for the terrace or patio, where they receive more warmth than in the open garden.

SELECTING THE BEST SITUATION

Try to pick a sunny spot for your tomatoes which is well sheltered from cooling and often damaging winds. This applies whether they are being grown in pots or beds. A warm sheltered corner is often the best choice. If necessary, due to constant winds, consider providing shelter by temporary fences of Netlon, Rokolene, a fence, or even old sacking over stakes or on rough wire supports.

PREPARING THE SITE

Tomatoes under glass and yielding a lot of fruit are a very greedy crop indeed. They therefore need a lot of sustenance. This is not so much the case when they are grown out of doors and crops are much smaller. Nevertheless they relish rich well-dug soil and good drainage. For border culture try to obtain some well-rotted farmyard manure (FYM) and dig this in at about 1 barrowload per 6–8 square yards. If FYM is difficult to obtain peat can be used to good effect in large quantities as a soil conditioner. Do not forget, however, that peat is acid in nature and that extra lime is needed to offset this.

Lime should be applied as a general rule at about 8 oz per square yard (producing a pH of 6·5, which is very slightly acid).

When there is doubt about soil conditions due either to physical structure or to potato eelworm (quite likely in an area where potatoes have been grown repeatedly for some years—see Chapter 10) it could be worth going to the trouble of laying down sheets of polythene and growing the tomatoes in ridges of soil or special compost (John Innes No 2 or soil-less composts). Trenches 9–12 inches deep, 3 feet apart, lined with polythene and filled with compost can also be used.

Alternatively, bolsters (see Chapter 4) are very useful for outdoor culture under doubtful soil conditions. Bolsters can be bought complete with compost. All that is necessary is to remove the top portion of the polythene to take the plants. Pots or containers filled with John Innes No 2 compost can also be used.

It must be appreciated, however, that tomato plants in containers of any kind, including bolsters, need constant watering.

PLANTING PREPARATIONS IN BORDERS

After digging the ground and applying farmyard manure or peat and lime, the ground should be well forked and firmed up with the feet. It is neither necessary nor advisable to rake it down to a fine condition. Three or four days before planting, evenly scatter a medium potash tomato base at the rate of 6–8 oz per square yard or 3–4 oz per yard of bed 18 inches wide. Where special compost is being used no base is necessary.

PLANTING

Recommended distances for outdoor planting are in rows 30–36 inches apart with the plants 15–18 inches apart. Try to run the rows north/south since this allows better sun penetration. Where only a few tomato plants are being grown, space the plants 20–24 inches apart, preferably in a south-facing border in front of a wall. Containers should be placed in the best position and spacing is not usually a problem. Bolsters can be run in rows 30–36 inches apart, as for border planting.

TIME OF PLANTING AND PROPAGATION

It takes much less time to propagate tomatoes in the spring than in the winter, due to the better light conditions and higher average temperatures. Planting times vary according to region, but on average mid-May planting is possible in the most southern areas. It can be earlier in places such as the Channel Islands. Usually it is better to wait until June in the north. Remember to use the growth made by early potatoes as a guide to planting time. Plants can generally be propagated in 4–6 weeks. This means sowing in early–mid-April.

Propagation under heated glass is preferable since more balanced plants are produced. The seed is sown and pricked-off as described in Chapter 5. There is much to be said for sowing directly into pots. Peat pots of 4¼-inch size are preferable, using a good compost. This enables plants to be grown to a fair size without detriment under glass or frames before planting out, feeding with liquid fertilizer when necessary. It is a good idea to allow the first truss to be in flower before planting. Ideally the plants should be moved from the greenhouse to cold frames. Here they are given progressively more ventilation for some 2–3 weeks before being set out. Alternatively, put plants in a sheltered corner for a week or so.

If plants are to be grown in cloches, frames or polythene tunnels, it is not so important to acclimatize them. Under such cover planting out can be effected much earlier, 2–3 weeks earlier on average, much the same as if grown in cold greenhouses. Always make sure that the cloches or frames are in position some 10–14 days before planting.

PLANTING

Planting should be carried out by carefully knocking the plants out of their pots, unless they are in peat pots or soil blocks. Now set firmly with the root-ball about 1 inch below the soil surface. A light watering-in is usually advisable unless the soil is wet. Using containers of any kind the procedure is the same, although watering-in is generally essential.

ESTABLISHMENT

If the plants are in good condition and have not been allowed to get pot bound (the roots in a solid mass in the pot) perhaps because of delayed planting due to weather, they should start to grow fairly quickly. Pot-bound plants take longer to establish. Should cold weather set in all plants can, however, 'sit' for some time. It will also be noticed that they usually change colour to a blue-green. Some temporary shelter from cold winds will help in this respect.

The foliage of outdoor plants invariably has a harder texture than that of indoor plants. The leaves will also frequently curl upward due to cold nights. Spraying the plants lightly with water will generally help to get them quickly established, particularly if the weather is hot and sunny.

SUPPORT

All plants except dwarf 'bush' varieties should each be given a 4½–5-foot cane either inserted vertically or obliquely at an angle of 45°. When plants are grown in cloches or elevated frames it is wise to run a wire horizontally on sticks 4–6 inches below the top of the cloche or frame to which the plants are secured and twisted along the wire.

SEASONAL CULTURE

Regular watering and feeding is essential throughout the whole growing season. Regular tying of the plants to canes or other support is also essential, using soft twine or wire paper-clips. Removal of side shoots is carried out as for greenhouse tomatoes, the plants being restricted to a single stem (except for bush varieties). Some defoliation will also be required, but not with the same severity as for greenhouse tomatoes. Only remove the lower leaves when they show signs of yellowing or rotting. A mulch of straw or peat helps to avoid damage to fruit which trails on the ground, as happens with bush varieties or those trained obliquely.

Plants in bolsters or containers must be watered and fed with great regularity otherwise they can quickly dry out. Spraying regularly with a fine rose on a watering can or hose-pipe helps the plants to set their fruits, especially in very hot weather. Medium potash feeds are the rule throughout the season, unless growth is either especially hard or soft. These can be either in diluted liquid form or applied dry and flushed in. Feeds should be applied at 14-day intervals.

Soft plants can be given sulphate of potash either in solution (1 oz per gallon) or a pinch per plant. Hard plants can be given dried blood (2 oz per gallon) or scattered on dry. Sulphate of ammonia can be used at the same rates.

According to the type of season the plants are 'pinched' out above the 3rd–5th truss, since it is usual to be able to ripen only 3–5 trusses (4 on average). A decision has usually to be made by August, according to the development the fruit is making. In cold areas or in bad seasons only 1–2 trusses may be possible.

Under the protection of cloches or frames results can be almost as good as with greenhouse culture, apart from the restriction in space. Some shading with lime and water may be necessary in very hot weather.

BUSH TOMATOES

Bush tomatoes are not kept to a single stem, being allowed to spread over the ground on a mulch of straw or wire netting supported 3–4 inches above ground level. Bush tomatoes are particularly useful for cloches and frames. Some restriction may be required, pinching out the leading shoots when they have used up their quota of available space.

At the end of the season the plants are removed and all unripened fruit lifted into a greenhouse or put on a light window sill to ripen.

The same pests, diseases and viruses affect outdoor tomatoes as tomatoes grown in greenhouses and these are dealt with in Chapter 10.

12
Soil Sterilization

Sterilization of a soil or growing medium is a term frequently used in gardening. Yet it is a misleading statement since a completely sterile soil would be infertile. The term really means the killing off of harmful organisms which includes pests, diseases or other forms of soil life, leaving behind the beneficial organisms.

For the tomato grower using soilless media sterilization is not necessary since both sand and peat (the main ingredients) are generally free from harmful pests or organisms.

When John Innes compost is to be used the loam content should be sterilized, although it is often more practical to buy in the compost ready mixed.

The small quantities of soil required for compost can often be effectively sterilized by the use of an electric sterilizer. On a small scale, wet soil can be spread on a sheet of corrugated iron supported on bricks over a brush fire. The soil should be turned frequently when it is steaming vigorously for a period of 10–15 minutes. Boiling water (or as hot as possible) poured over a shallow layer of soil on a clean surface, then covered with clean sacks to keep in the heat is also remarkably effective. The low-pressure sterilizers recommended by the John Innes Horticultural Institution are no longer thought to be very effective. These involve the building of a brick sterilizer composed of a shallow trough of water over which the soil is placed on a perforated steel tray.

The sterilization of soil by steam is by far the most effective method and is widely used in growing commercial tomato crops. Unfortunately, it is obviously not practical for the gardener.

Chemical sterilization is much more feasible. Formaldehyde is used at the rate of 1 part formaldehyde to 49 parts of water, or cresylic acid at the rate of 1 part cresylic acid to 39 parts of water. Both are applied at the rate of 5 gallons per square yard of the diluted solution and the soil is covered with sacks to keep in the fumes. A period of 4–6 weeks must be allowed to elapse before the soil can be used. There are definite limitations in the effect of the two chemicals.

A powdered chemical based on metham sodium (Basamid) may be available in small quantities and it is highly effective for border sterilization. It is scattered on and watered well into the soil, which is sealed with water. The soil must be above about 49 °F (9 °C) and approximately 40 days must elapse before the soil can be used for plants.

13
Tomato Varieties

There are a great many tomato varieties available today, and all have varying characteristics of growth, quality, shape of fruit, and in some cases resistance to disease, especially to leaf mould disease (Cladosporium) and to tomato mosaic virus (TMV). Many modern varieties are F1 hybrids—the seed must be produced annually under controlled conditions.

The following symbols refer to approximate vigour group, which includes the general nature of growth, but modern plant breeders can quickly change the habit of growth.

T/S = Tall spreading
I = Intermediate
C = Compact

Those marked with * are resistant to various strains of leaf mould (Cladosporium) disease (A B strain or both).
† Varieties bred by Glasshouse Crops Research Institute.

IMPORTANT NOTE

TOMATO VARIETIES

The following is a list representative of those varieties available offered by the three major seed houses—Asmer Seeds, Leicester, D. T. Brown, Poulton-le-Fylde, Blackpool, and Suttons Seeds Ltd, Reading, and it will be noted that the main concentration is on F1 hybrid varieties and that many of the 'straight' varieties are not listed. While it is appreciated that many gardeners may well wish to obtain varieties not listed it must be pointed out that there is a tendency in the British seed trade to reduce the number of 'straight' varieties in keeping with general EEC seed trade policies.

F1 Hybrids

*Ailresist** Ailsa Craig type. Good for poor light areas (T/S).

*Amberley Cross**† Early heavy cropper. Greenback free (T/S).

*Arasta** Early non-greenback. Strong compact habit. Good yielder (C).

*Asix Cross** Heavy and early cropper (C).

*Bonset** Early cropper. Verticillium resistant (I).

*Clavito** Semi-greenback, short jointed Autumn Crolpury (I).

*Eurobrid** Non-greenback. Early. Resistant to fusarium and verticillium.

*Eurocross A** Non-greenback fruit. Excellent colour and quality (I).

*Eurocross B** Large round fruits. Excellent for early forcing on heavy soils (I).

*Eurocross BB** Large multi-locular fruit. Very early heavy cropper (I).

*Extase** Short jointed heavy cropper. Semi-greenback fruits of high quality (C).

*Cudlow Cross**† Heavy cropper (I).

*Findon Cross**† Early (C).

*Florissant** Vigorous. Yield and quality excellent (T/S).

*Fontwell Cross**† Early (T/S).

*Globeset** Ailsa Craig type for heated or cold greenhouses (T/S).

*Growers Pride** Early and vigorous (T/S).

*'Harrisons' Syston Cross** Short jointed free setting type. Fruit of good quality and colour. Good for cool greenhouses (C).

*Hollandbrid (C118)** Non-greenback. Early. Resistant to fusarium and verticillium (I).

*Ijsselcross** Non-greenback. Vigorous with larger fruit than Maascross. Moneymaker type (T/S).

*Kingley Cross**† Non-greenback. Short jointed. High yielding (C).

*Kirdford Cross**† Compact form of Selsey Cross with TMV resistance (C).

*Lavant Cross**† Early (T/S).

*Maascross** Non-greenback. Very early cropper with excellent quality fruit. Comparable to Eurocross B but more vigorous (T/S).

*M.G.** Ailsa Craig type (I).

M.M. (and other strains—Super, Nova, etc.)* Moneymaker type (T/S).

*Moneyglobe** Excellent quality. High yield. Free setter and earlier than Moneymaker (I).

*Pagham Cross**† Sustained cropper (T/S).

*Primset** Non-greenback. Very early. Good yield (C).

*Plusresist** Early semi-greenback. Suitable for heated or cold greenhouses (I).

*Rijncross** Semi-greenback. Vigorous with well-shaped fruit (T/S).

*Sarnia Cross** Very early non-greenback. Resistant to verticillium and fusarium wilts (T/S).

*Selsey Cross**† Non-greenback. Heavy cropper (T/S).

Seville Cross (Suttons) Non-greenback. Early cropping (I).

*Seriva** More vigorous and heavier cropper than Moneymaker. Large fruit. Non-greenback. TMV tolerant (I).

Sorrento Cross (Suttons) Non-greenback. Heavy cropper (I).

*Supercross** Non-greenback. Fruit similar to that of Moneymaker. Tolerant to TMV (I).

*Topcross** Suitable for growing in heated or unheated greenhouses. Excellent fruit quality. Very good yielder (I).

*Virocross** Resembles Supercross but semi-greenback slightly larger fruit. Tolerant to TMV (I).

*Warecross** Good quality. Early. Very high yielder (T/S).

Straight Varieties

Alicante (Suttons) Non-greenback. Excellent quality. Early (T/S).

Ailsa Craig Medium-sized well-shaped fruit (T/S). Many strains available.

Best of All (Suttons) Large cropper. High quality (T/S).
Craigella Ailsa Craig type. Non-greenback (T/S).
Dutch Victory Non-greenback. Grown cold in Holland for the English market (I).
E.S.1 Early cropper. Ailsa Craig type (T/S).
E.S.5 Heavy cropper. Fruit of excellent colour and quality (T/S).
Exhibition (Stonor) Heavy cropper. Even-fruited (I).
Fortuna Potato leaved—for outdoor cropping (C).
Harrison's First in the Field Good cropper. Early (outdoor only) (C).
*J.R.6** Short jointed. Free setting (C).
Minimonk Moneymaker but of compact habit (C).
Moneymaker (Stonor) Very popular. Non-greenback type of medium size and good cropper. Well-shaped fruit of good colour (I).
Potentate Very heavy early cropper (I).
The Amateur Bush variety. Suitable for cloches. Early.

Yellow Varieties

Golden Dawn Early, medium-sized even-shaped fruit on strong trusses.
Mid-day Sun Golden Orange, medium-sized fruit. Does well outdoors. Vigorous.

Pelleted seed can be obtained of certain varieties to facilitate sowing.
Root stock seed of KN, KVF and KVNF is available.

Fruits of different colours or striped are available from Practical Plant Genetics, 18 Harsfold Road, Rustington, Littlehampton, Sussex.

Index